NATURE
LOVE
MEDICINE

ESSAYS ON WILDNESS AND WELLNESS

NATURE
LOVE
MEDICINE

ESSAYS ON WILDNESS AND WELLNESS

Edited By
Thomas Lowe Fleischner

TORREY HOUSE PRESS

SALT LAKE CITY • TORREY

First Torrey House Press Edition, November 2017
Copyright © 2017

Published by Torrey House Press
www.torreyhouse.org

International Standard Book Number: 978-1-937226-77-0
E-book International Standard Book Number: 978-1-937226-78-7
Library of Congress Control Number: 2016943392

For my parents: Rose and Warren, who raised me to recognize the beauty of this world. And for my children: River and Kestrel, who constantly rekindle my faith in the future.

—TLF

Contents

Entry

This book unabashedly declares that love for our biologically and culturally diverse world is a fundamental virtue and imperative. Humans are born to practice natural history—simply put, to pay attention to nature—and are physically, emotionally, and socially healthier when we do. Collectively, the contributors explore this notion that attentiveness to nature remains an essential pathway to sanity and health, for both individuals and societies.

The twenty-three writers gathered here address three interrelated questions, yet follow distinctly different paths toward answers. Ecologists, psychologists, and educators; poets, novelists, and artists; researchers and creators; theorists and practitioners; the healers and the healed ask: how can nature heal us, as individuals and communities? How does practicing natural history lead to healthier lives? And, why does deep engagement with nature help us love the world? From different nations, cultures, professions, religious traditions, and approaches to engaging the world, these writers explore how natural history—a practice of intentional, focused attentiveness and receptivity to the more-than-human world, guided by honesty and accuracy—is, at its core, the practice of falling in love with the world.

When we pay attention, we connect. When we connect, we become more whole. The words *whole* and *heal* derive from the same Old English and Proto-Germanic roots; for at least seven hundred years our language has recognized that wholeness and healing sprout from the same source.

So—natural history: a practice that leads us toward wholeness and healing as we make our way within this world shot through with wildness, that ineffable wisp of life force, the sinew that braids us together.

Nature. Love. Medicine. Healing ourselves and the world.

—TLF

Nature
Love
Medicine

The Gods Are Not Large

JANE HIRSHFIELD

But perhaps
the heart
does not want
to be understood.
Your shadow
falls on its pond
and the small fish
hurry away.
They have
their own lives,
not yours,
which they love.

And if to you
it is anger,
bewilderment,
grief,
to them
it is simply life:
their mouths

open and close,
their gills,
they are fed,
they breathe.

The gods
are not large,
outside us.
They are the fish,
going on
with their own concerns.

Our Deepest Affinity

THOMAS LOWE FLEISCHNER

Still dark. We crouch low, careful not to wobble the narrow dugout canoe, wary of the Black Caimans known to lurk in the water of this lagoon, off a side-channel of a tributary of the Rio Napo. A few miles downstream this major branch of the Amazon will become the border between Ecuador and Peru. As daylight gradually suffuses the Amazonas dawn, six of us—four travelers led by two young Kichwa guides—silently skim across the lake's surface. Grays imperceptibly shift into greens. Loud squawks and whistles signal the hidden lives of birds, emphatically calling in a new day. We cross the length of the lagoon in twenty minutes, then turn up a narrow black-water stream. Tiger-Herons stride the dank sandbars. At this low-water season, the canoe bumps into downed logs, and we grab the gunnels to steady ourselves.

We arrive at an indistinct trail through the forest, clamber out of the canoe back on to terra firma, and begin walking—away from water, deeper into this "green lung of the world." This is not my native habitat—I lean toward open deserts, stone canyons, and windy alpine reaches—but I'm awestruck by this diversity, this clamor of life upon life. Whichever way I turn my head, I observe a dozen or more types of trees, most unknown to me. Each flash of wings is likely that of a new species of bird or butterfly: wafting butterfly flight sparkles azure in the glinting sunlight; manakins, gold and black, flick from shadow to shadow. A troop of diminutive Squirrel Monkeys leap through the branches above us, babies clinging to their backs, while we hear the far-off roar of Howler Monkeys, a sound strangely akin to distant aircraft engines.

Finally, we reach an enormous Kapok tree—the most massive plant I've ever seen in the tropics. It would take a dozen people clasping hands to encircle this tree, which stands well over two hundred feet tall, with a straight, clean trunk. The immense erect form of this giant would not be out of place in a Pacific Northwest conifer forest, or a California redwood grove. The local Kichwa community has built an immense ladderway up into the green. Our guide's father and uncle were among those who climbed, unroped, to scout the canopy when the local community first determined to build this elevated window on the forest.

We ascend the two hundred stairs, and then step out onto the broad wooden planks—truly, the highest tree house I've ever seen. From this vantage point, near the top of the forest's tallest tree, we can look *out* upon something usually not visible, the broad unbroken expanse of the rainforest, and *down*, onto the canopy, the roof of this forest community. Enormous pale branches curl dozens of meters outward toward the gauzy-blue morning sky, as if my own arms extended out to embrace the world. Leaves of every green hue and a hundred different shapes and ruffled textures unify into a billowing surface pulsing with life.

Here, we indulge our eyes and hearts for the next two hours—a parade of birds arriving, departing, sometimes landing on the very branch that cradles our airy platform. In this pair of hours, I see twenty-two new species of birds, all of them astoundingly beautiful. I'm not the sort of birdwatcher that fixates on numbers or tallies, but still, the flood of new sensations—new colors, sizes, shapes; whole new families of birds—*electrifies* me, as if sparks flash from the perimeter of my body, mix with the pulsating energy of this greatest of Earth's forests.

Just below us, on the next tree, two toucans clap their outrageously large and cumbersome bills—almost as large as their bodies, like polished, gilt-edged sheaths for giant curving daggers. An owl-like Great Potoo sits stock-still, staring at us from our own branch, while a nunbird—black body, curving golden bill—perches upright for long minutes from the next tree. The iridescent blue-violet head and luminescent golden legs of a Purple Honeycreeper flit between glossy leaves of the tree immediately below me. Others of the exquisite tanager tribe—Silver-beaked, Turquoise, Opal-rumped, Masked Crimson—also dazzle us, revealing shockingly vivid colors—scarlet, azure, saffron—as they methodically probe branch crevices and leaf surfaces for insects.

Suddenly, shrieking with sociality, a dozen *Amazona* parrots, each as long as my arm, fly close by, well below us, and disappear into the humid void.

Who knew that angels' backs glistened metallic green?

Or that we, mere bipeds, could ever observe them from above?

Another season, another place.

The crimson bird perches, almost motionless, on the stout horizontal branch of the Velvet Ash, twenty feet above my head. Every few seconds, its substantial bill opens slightly as it chirps loudly—*pitúk*. For almost ten minutes, I quietly watch and listen to this repetitive behavior. The only other sound and motion: wind tousling these oasis trees, ruffling the surface of the deep green pool, beneath the abruptly rising russet limestone cliffs.

My daughter, soothed by the day, just finished with a year of college, dozes peacefully on the sand, a few feet behind me.

It's no accident she and I are here together in this moment, able to witness this Summer Tanager, far northern outlier of a group of wonderfully colorful tropical birds, seemingly out of place in North America's desert Southwest.

It had been a hard week—stresses at work that swallowed days, created unexpected meetings, and traced shadowy outlines of loss in the lives of friends. Losses of jobs and identities, and, in one especially tragic case, the loss of life, when a colleague, riding his bike home from work through the desert, was abruptly smashed to the pavement and sudden death by a drunk driver.

All this—convening coworkers, responding to somber concerns, explaining decisions and actions—kept me tethered to a computer screen, the flat face of illusion and anonymity. On my way home from a culminating, especially draining meeting, bruised by the challenges and sorrows within my community, I realized my spirit needed rejuvenation. I needed *beauty*. So I proposed to my daughter—a healing balm in her own right—that we hike together into one of my favorite canyons the following day.

We drive through the parched tan landscapes of upland Arizona in June, the edgy time of year when subtropical moisture amasses into towering clouds, but monsoon rains—that for which generations of rain-dances were danced—taunt us with their nonappearance. We cross the crest of Mingus Mountain, passing through cool pine forests, then descend into the heat of the valley on the steep downside of this fault-block mountain. We swish and sway down the curves of the two-lane road, finally turn off and cross the green river, true to its name—Verde—and then veer onto ten miles of bumpy dirt.

We scramble out at the trailhead, and descend quickly toward the lush beauty of a spring-fed wilderness canyon. From one moment to the next, as we arrive at the canyon bottom, we leave behind the heat and the dust, and are enveloped by deep shade, and the shifting burble of moving water.

In we walk, enfolded within shimmering green, with red cliffs at the perimeter of vision, bolstered by the promise of watery escape. This new world is rich with birdsong and tantalizing flicks of yellow, red, and blue amongst the fluttering leaves.

Whenever we feel like it, we stop and sit quietly at the edge of this stream, rushing with this great arid-lands miracle: abundant clear water in motion. Like the stream, our conversations meander—new job, new relationship, upcoming travels. Like the smooth sandstone ledge we sit upon, our minds begin to still.

What I never doubted, nor even had to consider: that coming here would heal us. That the cradling embrace of this bio-diverse canyon would be good medicine.

🌾 🌾 🌾

Our world desperately needs more people to be more in love with it. Anyone reading a newspaper, browsing Facebook, searching the Internet, or, sometimes, just walking down the street, knows we live in a challenging time. We're routinely confronted with conspicuous anxiety, depression, violence, with evidence and news of climate change and terrorism. Too often, people conspicuously mistreat each other, and blatantly disrespect the natural world.

But the Earth is a gift, not a problem—and loving the world is as important as grieving for it. Being an awake and engaged human being is rooted fundamentally in an unequivocal love of the world. Our deepest affinity is for this rich and remarkable world we live in—our fellow beings, the textures and colors of landforms, the luscious scents of each place we touch. The term "biophilia" indicates our innate urge to affiliate with other forms of life.

As we fall more deeply in love with the world, we learn to love ourselves more fully, and learn to care for our surroundings and our fellow beings more completely. Paying attention to nature—something beyond ourselves—is how we develop, nurture, and express our love of the creation. All good things emanate from this affection for the world—caring, compassion, deep engagement, and meaning. So it behooves us to consider: what fosters passion for the planet?

The attentive practice of natural history—which I've previously defined as "a practice of intentional, focused attentiveness and receptivity to the more-than-human world, guided by honesty and accuracy" is, essentially, the practice of falling in love with the world. Such practice can involve watching birds at a backyard feeder or in alpine tundra; tending the garden or observing old growth forests; sketching flowers, watching butterflies, fly-fishing, or counting migrant hawks. In short, natural history involves mindful observation.

Natural history, then, is a verb, not a noun—a *practice,*

something we *do*. And attention, as the Buddhist psychologist John Tarrant has pointed out, is the most basic form of love. Human beings are designed by natural selection to do natural history—our senses, our limbs, our whole bodies have evolved for attentiveness. Natural history is humanity's oldest continuous endeavor. No wonder—our survival has wholly depended on our capacity to pay attention to the encompassing living world, full of threats, foods, and delights.

Yet, we live in a very odd historical moment: there has never been a time in the history of our species when so few of us have paid attention to the world that surrounds us. The current gush of social dysfunctions—violence, depression, anxiety, alienation, lack of *health* in so many ways—coincides with the mass sacrifice of human interaction with nature, the greatest dearth of natural history in human history. We have come to see the world as a funhouse built of human mirrors, where we see only ourselves, and narcissistic distortions of ourselves. Without attentive immersion in the larger-than-human world—the exact immersion for which we are biologically adapted—we dissolve into individual and collective malaise.

Natural history combats arrogance, tackles despair. Attentiveness to the world around us engenders humility and open-mindedness. It humanizes and grounds us by offering a larger perspective on the world. Humility—so sorely needed in our social and political discourses—can only be taught through modeling. Some human elders do a fine job of this. But immersion in the complexity, unpredictability, and, occasionally, the ferociousness of the natural world almost always teaches humility.

Natural history makes us healthier as individuals and, collectively, as societies. Engagement with the beauty and power of Nature brings out our best behavior, supports our best selves.

The expansive attentiveness of natural history is ultimately hopeful. For we're all wired to pay attention to nature. Human consciousness developed in natural history's forge—our patterns of attention were sharpened as we watched for danger and sought food. Practicing natural history is our natural inclination. We are all born with the capacity to attend the world with wonder—just watch any child—and so recovering a collective sense of stewardship is within our grasp.

When we love the world we can love ourselves. And vice versa.

Each moment we pay attention reverberates with heightened perception, as awareness from the past intersects with that of the present.

Just now, a tiny pale butterfly tumbled past in streamside flight, and I caught the briefest glimpse of orange at the outer margin of the wings. Based on many past observations, I recognized this as a Desert Orangetip. Instantly, based on past study, a notion of this lustrous insect's life enters my mind: what it eats, where it will lay its eggs, how long it will live, that it can persist as a chrysalis for several years, that its caterpillars eat wild mustards, that its adults gather in fluttering hilltop congregations.

One quick burst of quivering white, laced with orange, and I'm connected with the here and now of communion, and the practice of empathy.

The more we pay attention, the more we find beauty. Natural history opens up entire worlds: the remarkable golden hairs on the anthers of a *Penstemon* flower; the graceful tilting flight of a harrier just above the field where it hunts; the glossy green sheen on a tiger beetle's carapace; the bold rings of orange, black, and white that encircle a Sonoran Mountain Kingsnake; the sudden flash of scarlet when a male blackbird displays its wing patch; the lustrous surface of sandstone traced by our fingertips. The facets of beauty available for us to notice are, truly, beyond counting.

It's easy in our troubled times to think that something as simple or abstract as beauty should be shunted to the sidelines, that it is overly self-indulgent to care about the beauty of this world. Or to remember that we humans are lovely, fragile, flawed—but, still, beautiful beings. But I concur with *The Idiot*. The title character in Dostoyevsky's novel declared, "beauty will save the world." As an evolutionary biologist, I'm often struck by the implausible reality that the world is more beautiful than it needs to be. We could have a functioning "system" with fewer parts, less extravagance, perhaps more efficiency.

But each day we start anew and walk out into a world that is full of sorrow and injustice, yes—but that is also heartbreakingly beautiful. Despair must not overrun our appreciation for this world—plant and animal, stone and sky, and human souls—that is immeasurably lovely, more beautiful than it needs to be, in spite of the grief that is embedded within it. As Terry Tempest Williams asserts, "beauty is not a luxury but a strategy for survival."

Our capacity for honoring beauty is interwoven with our aptitude for compassion. Compassion—literally, *feeling with*—occurs when we encounter others with open hearts. It has often been noted that we are less able to respond to human suffering when it is on a massive, impersonal scale—tales of far-off flooding, yet another shooting, or even genocide. But when a story becomes individual, our psyches can respond more compassionately to *this* victim, *this* survivor, *this* neighbor in need. It works the same way when encountering our non-human neighbors. When we connect with individual, specific lives—*this* flower blooming in this parched mudflat, these muskrat eyes looking back at me from the desert pool—we can transcend nebulous notions like "Nature" and replace them with texture, depth, and a realm of specificities. Natural history, then, is a path of compassion.

Of course, there's no shortage of wonderfully selfish reasons to care about nature—because it's good for us.

"Nature Deficit Disorder," a term coined by Richard Louv, has focused public attention on the notion that immersion in nature is essential for human health. The idea garnered some persuasive power by utilizing the jargon of medical science— "deficit," "disorder." As yet, there's no formal medical disorder, nothing listed in the *Physician's Desk Reference*, concerning lack of nature. But perhaps there should be. When one peels back the guard hairs of cognitive science and starts reading what's buried in technical journal articles, it's stunning how remarkably *healthy* time outdoors turns out to be. Again and again, the conclusion in a wide variety of psychological and medical studies—published in such respected journals as the *Proceedings of the National Academy of Science, Journal of Experimental Psychology, Environmental Health and Preventative Medicine, Journal of Cardiology*, and many more—is that it's simply healthier, both physically and emotionally, to spend time outdoors than in more human-dominated urban settings, to walk in forests rather than along city streets.

The documented benefits of what healthcare professionals now sometimes refer to as "nature therapy" or "forest therapy"—simply put, being outside—are striking, and diverse. People who spend time outside have less stress, improved memory, stronger immune systems, better vision, increased creativity and problem-solving capacities, reduced inflammation, improved concentration, and are, well, happier.

Immersion in nature reduces stress, and this can be measured by standard physiological metrics like heart rate, heart rate variability, and blood pressure. A Harvard study demonstrated a direct linkage between exposure to green spaces and mortality rates: of the more than 100,000 female nurses monitored, those living in the "greenest" areas had a twelve percent lower mortality rate than those living in human-dominated habitats, and cancer risk declined even more. In other words, living closer to nature can help us live not only better, but longer.

Several studies have demonstrated that urban people living near parks or greenspace have lower levels of incidence of psychological problems, and that people who visit such greenspaces have lower levels of stress hormones than their peers who had not been outside recently. One study concluded that "stressful states can be relieved by forest therapy." In another, people who walked in parks showed lowered blood flow to the part of the brain where brooding ("morbid rumination," to the psychologist) takes place. The researchers concluded that getting out into natural environments could be an easy and almost immediate way to improve moods of city dwellers. In another study, a controlled experiment revealed a direct relationship between separation from nature and increase in risk factors for mental illness. These researchers concluded that "accessible natural areas may be vital for mental health in our rapidly urbanizing world."

A study in Toronto quantified the health benefits of living in neighborhoods with a higher density of trees: having eleven more trees on a city block decreased cardio-metabolic conditions (heart disease, hypertension, high cholesterol, stroke, and so on), in ways comparable to a boost in annual personal income of $20,000, or being a year and a half younger. Living with trees, one could say, makes a person richer and more youthful. A large body of research, from several nations and continents, has documented that outdoor activity in children reduces their risk of nearsightedness as adults. Playing outside, in other words, literally helps us see the world more clearly. Time outdoors has been correlated with boosted immune systems in general, and there are indications that forest environments may stimulate production of anti-cancer proteins.

These are but a few snapshots of the exploding scientific literature on the healing power of nature. The bottom line: nature is good for us. This is no longer just the murmurings of nature poets or the trumpetings of conservationists, but the prescription from a wide variety of cutting-edge, peer-reviewed studies in cognitive and medical sciences.

🌿　🌿　🌿

It's the final night of a week-long group backpack into red earth. After a late supper and lingering conversation, two of us lay back, stunned into horizontality by the night sky. In this canyon of smooth stone, deep in the heart of one of the largest roadless areas in North America, our eyes gaze intently where they rarely focus—upward. The dome of stars is phenomenally textured, with vast depth visible within the Milky Way. Swirls and eddies of pulsating light contrast with the velvet black of infinity. Arcing downward, the gaudiest shooting star I've ever seen blazes brilliantly for a full twenty seconds as it angles toward Earth. Its long tail remains gleaming even after I close my eyes.

It's a brief interlude of extended psyche, as our minds momentarily stretch beyond their usual boundaries—a sense of delighted union with the larger forces of not only this world of green, but the universe that surrounds it. And better for sharing it with a close friend.

Lying, face-up to the heavens—dazzled, humbled, entranced by the endless array of light—provokes a fundamental feeling of kinship with people through times and cultures. We all look up, and wonder.

And on these sandstone slabs, in front of this long-inhabited alcove, near thousand-year-old petroglyphs and discarded clay pots, we see the same framing of the universe as our ancestors—unchanged through human time, but for the occasional steady scoot of a tiny satellite's light.

In these singular moments, we are the outbreath of galaxies.

The Silence of the Forest as Our Lover

GARY PAUL NABHAN

I.

...the silence of the forest is my bride and the sweet dark warmth of the whole world is my love and out of the heart of that dark warmth comes the secret that is heard only in silence, but it is the root of all secrets that are whispered by all the lovers in the beds all over the world.

Thomas Merton (1997), *Dancing in the Water of Life*

Among the earliest memories imprinted in my mind: Sitting alone in the sands of the Indiana Dunes when I was three, maybe four years old. Listening.

The late afternoon sun was cascading down through the canopies of oaks and cottonwoods above me. A squabble of Blue Jays appeared to be my only companions for well over an hour. I became mesmerized by their presences.

I sat in my corduroy overalls, my bare feet slightly buried in the sand to ground me, my ears wide open to catch the sounds all around me, and that is how I heard the jays converse.

While the sun slowly set, I sat in rapture as they sang to me about the intimate workings of the world that could be found both around and within us.

Suddenly, another voice broke my trance. It was my mother's voice, beckoning me to get up and come into the house:

"G.P.! Jeepee! Jeepers Creepers, get on in here, it's dinner time."

I could barely follow her command, for I was still lost in another conversation.

"Jeepers Creepers! Quit your daydreaming! Get off your little butt and come in here before it's too dark for you to do any more daydreaming!"

I was being jolted out of my reverie and I was not yet ready to leave it behind:

"Shhhhhhhh! Mama, the jays are still talking to me!"

"Look here, you little rascal. It's your mama talking to you, not some Blue Jays. Now get on in here right now before I give you a spanking!"

I moaned. "Mama, didn't you just hear her? She really didn't like that you said that!"

"Do you want dinner or a spanking? Those are your only two choices, Buster. Quit this nonsense and get into the kitchen!"

I was suddenly so grief-stricken by the pressure to choose between the human household and the wild kind that I could not keep from crying.

"Mama, don't you ever say that again, or the jays will come after you!"

I ran inside the house and hid under the bed, refusing to eat or to talk to anyone. Heart broken, my rapture dissolved into despair and divided consciousness.

Perhaps that is how we fall from the garden, fall out of love and fall into a darkening cloud of doubt and abstractions. Suddenly, the birthright of being intimate and in love with the world is challenged, and it takes both time and other kinds of love to recapture that rapture.

II.

Of course, love and intimacy eventually come around again. I believe I was five when a little girl from Pittsburgh walked into my life. Her father rented a house in the Indiana Dunes for just the summer. I have no idea how we found one another and I can't even remember her name. No matter. Words were not much of a deal between us. We spent most of our time together holding hands, walking in the oak woodlands that held the sand in place on the oldest, tallest dunes.

In retrospect, I can't remember any time that we spent inside together. After I had served my sentence in kindergarten, my mother felt sympathy for me. She helped me take off my ill-fitting shoes, threw them away and told me to go play. I found this new friend down the street and neither her mother nor mine worried at all if we trekked into the nearest woods for an hour or two each morning.

In the glare of a hot summer morning, we would crawl barefooted and barehanded up the steep slope of a big dune. When we got to the top of it, we would disappear into the cool, dark shade of the forest and follow deer trails across a dunescape the size of a football stadium. Because I had already explored the trails with my older brother, I became the native guide, escorting her to see and smell the hidden wonders of the wilderness.

I can only recall one or the other of us pointing to things, then feeling their textures, smelling their fragrances or tasting their flavors. We picked flowers and drank their honey-like nectar. We smelled the sap on the trunks of jack pines. We dipped our fingers into the tannin-rich water pooling in a rotting oak trunk, then sucked on our own fingers to discern whether it tasted like root beer, as the roots of sassafras did. We ate little raisin-like morsels found under bushes, which I later learned were rabbit droppings from my horrified Aunt Jeannie.

Near the end of the summer, on what would become our last time in the forest together, we both squeezed into the lightning-struck hollow of an old sycamore. I was so close to her, I heard her heart beating. Her right hand, still held in mine, seemed to be hot to the touch, as if she had just warmed it in front of a campfire where we had come together to roast marshmallows. I could not be sure whether I was smelling her essence or that of the musky fungi growing out of the wounded heartwood.

I cannot remember her name, but I am still in love with that fragrance.

We squeezed out of the tree and walked back into the bright light of the day.

Back at my family's house, we stumbled upon a garter snake resting halfway up the flagstone stairway that reached to the top of the dune right at our front door. We did not dare to cross it, so we came up to our house from the backyard.

We each stood on a concrete block wall on opposite sides of the stairs, looking down at the snake curled up in the shade, saying nothing. We did not know snakes to be dangerous, certainly not phallic, let alone evil. Its deep greens and yellows were simply beautiful. Who would not be placed in awe by the mere sight of such a creature?

Then it uncoiled, and slithered up the stairs between us. It startled me and I lost my footing, falling toward the snake itself.

My head hit a flagstone stair just above my right eye.

I remember getting up and running toward the house.

I never made it on my own, passing out before I reached the front door.

Later on, I learned that the little girl ran into the house to tell my mother that I was bleeding. In the confusion, she thought that I had hurt the snake or the snake had hurt me. She started to cry.

My mother grabbed the car keys, ran out the door, picked me up, and carried me to the car. She went back up the stairway and took the crying girl by the hand and placed her next to me in the front seat. She drove our '57 Chevy sedan down to a house on the next block and honked the horn to attract the attention of the little girl's mother. When the woman came to the car, my mother quickly explained why the girl was crying and said that she needed to rush me to a doctor's office two miles away.

The next day, my right eye was swollen shut. My face was black and blue. The doctor guessed that I had suffered a minor concussion. The girl was gone, the snake was gone, the summer was gone.

It was time to go back to school for the first grade—my sweet friend had already gone off to Pittsburgh—but I was too ashamed to go out into the world until I could see out of my right eye.

I did not recognize myself. I had lost my first love. To make matters worse, I had to get used to wearing shoes. For several more months, I could no longer feel the warmth of the dunes sand on my toes, smell the fragrance nor hear the heartbeat of the girl clinging to me inside the hollow tree.

Although I never saw her again, her fragrance would come back to me whenever I encountered the scent of blessed ferment in the dark warmth of a forest.

III.

When I was a teenager, a good part of my hometown was burned to the ground in the days following the murder of Martin Luther King, Jr. When the Democratic Convention was held in nearby Chicago, college-age brothers of my high school friends were knocked over by the force of firehoses, beaten with batons by policemen, or hand-cuffed and hauled off in paddy wagons to "drunk tanks" for protesting the Vietnam War. Mayor Richard Daley famously ordered his officers to "shoot to maim" if they found any protestors out in the streets after curfew.

I, too, protested the war by conspiring with friends to set off fire alarms in our high school, and I, too, spent a night in a drunk tank with friends, without ever having broken the law, just so a cop could "show us a thing or two." Fortunately for us, the cop was caught arresting us for being out after curfew before the curfew had actually begun. But the odors I sniffed in the drunk tank were not anything like the fragrance of the young girl who held me tight, hidden in the dark wood.

Under such social stresses, I dissociated from my family, from mainstream society, and from most of my old friends. The remaining ones—among the youngest members of the radical Students for a Democratic Society—participated in increasingly risky behavior that went far beyond non-violent civil disobedience.

They were angry young men hell-bent on inciting a revolution, while I was not angry so much as I was heartbroken—the world as I knew it seemed to be falling apart.

I still do not know what exactly triggered it—a virus or emotional stress from my parents moving toward divorce and my friends moving toward drugs and violence—but I had what they used to call "a breakdown." I tested positive for Epstein-Barr, but it was not from the infamous "kissing disease," mononucleosis. Perhaps it was what we now call "chronic fatigue" or fibromyalgia—we will never know—though I suffered a relapse of these symptoms a quarter-century later.

In short, my immune system tanked. I was too weak to go to school, and when I did, I vomited or passed out. I slept for days on end, waking up only to urinate. I could not hold down food, nor could I remember what day it was or why I was not in school. I was absent from the courses of my junior year for so many weeks that I was essentially written off as "a dropout." In fact, I never returned to enroll as a senior, and never received a high school degree.

But then, rather oddly and unexplainably, when springtime came, I asked my parents if I could go walk alone in the dunes to get some fresh air and exercise; perhaps being bedridden was what was holding back my recovery.

As the ducks and geese flew down their flyways from Canada to the southern shores of Lake Michigan, they, like me, found a wilderness stopover—a sanctuary of sorts—in the marshes and moors of the Indiana Dunes. I would go there alone, and spend hours listening to them honking and clucking. I waded out into the still cold water of the wetlands, rubbed the marsh's mud and slime all over my body, and hid in the cattails so I might see the waterfowl without them seeing me.

When I got hungry, I pulled up cattail stalks, peeled them like a scallion to rid them of their muddy film, and ate their crisp flesh. When cattails could not stave off my hunger I caught frogs and toads, snapped their legs off of them and ate the rare flesh around their leg bones like one might do barbecued spare ribs.

When summer came, I moved out of my parents' broken home and into a screened-in porch that sat by itself atop a sand dune overlooking Lake Michigan. It was on the property of one of my mother's oldest friends who lived right next to the house in which I was born.

I stayed there all summer—a hermit of sorts—making sketches of plants and animals, and reading. Early one morning, I rescued a roadkill Woodchuck, took it back to my picnic table, and sketched its image with pencils for forty-eight hours straight, until it began to stink. I borrowed from my elderly hostess a few of her books by Walt Whitman and Herman Melville and Rachel Carson. She directed me toward the dunes naturalists of her own era, who quickly became my heroes: the reclusive Diana of the Dunes; the world-famous wildlife tracker, Edwin Way Teale; and the lover of great American trees, Donald Culross Peattie. In fact, the first "nature book" I ever owned was Peattie's *Flora of the Indiana Dunes*.

But I did not peruse that fine local flora simply to "key out" and "name" the native plants that were flowering all around me. Instead I used it as a palimpsest, a rudder by which to steer me toward a safe harbor, a refuge. I was looking for secure anchorage in an otherwise uncertain if not tumultuous world.

That said, I did not think of myself as a naturalist, nor did I keep any notion that I was about to embark on a career in the natural sciences. I was simply trying to rid myself of heartbreak and sorrow by the only way I knew how: to go errant, to play hooky, and to leave the human world for another.

I suppose that I was endeavoring to heal myself by immersing my broken heart, mind, and body in the healing mud of the world.

By the end of that summer, I was turning toward the warmth and love of a massive, glowing energy field that I had largely ignored or dismissed since I had become a teenager. And yet, it was suddenly undoing and redoing who I was, and who "it" intended me to be.

Without even so much as a GED, I entered a small liberal arts college in the Midwest on probation, but within six months, I went truant, abandoning classes to work as a volunteer at the Washington, D.C. headquarters for the first Earth Day. I do not even recall how I learned that the first Earth Day was being planned for April 1970; I do not even remember how I got to the District of Columbia or found a place to live, for I was still seventeen at the time that I hastily made the decision to drop everything for the love and defense of Mother Earth.

I became a journalist-intern for the headquarters' news magazine, *Environmental Action*. At that moment in my life, I did not know how to write prose. I spent most of my waking hours writing poetry, drawing pen-and-ink sketches, or crafting cartoons. My "boss"—a civil rights and anti-war activist named Sam Love—didn't seem to know any more natural science than I did, but he somehow encouraged all of us to explore and express the connections we felt between "the environment" and "social justice."

We did not even know of the phrase "environmental justice" at that time, if it existed. Instead, we sensed that it was hiding somewhere in the woods, and would call out to us with its own grace note when it was time to do so.

A month after I turned eighteen, I was sent out from Earth Day headquarters on the East Coast to a small Christian college in the Midwest that overlooked the Mississippi River. I remember only four features of that first Earth Day: the spring rains were falling hard; the water level in the muddy Mississippi was close to surging over its banks and sand bags to soil everything in sight; I was the youngest speaker onstage that day; and I could not remember a single word that I had planned to say.

No matter. It wasn't about me anyway, it was about something that was calling *to* us and *through* us.

All I knew for sure was that spirited beings of the forests such as Blue Jays were still calling me, and I was certain that their voices mattered. What they beckoned me to do was stay connected to that vast, often silent but sometimes terrific power of love emanating from all the forests, deserts, scrublands, marshes, prairies, steppes, rivers, lakes, and oceans of this whole blasted planet, as we dizzily spin out to wherever destiny leads us to be.

They beckon you as well. Hold on tight, my friends, it's bound to be a wild ride!

Branching Out

NALINI NADKARNI

The attentive practice of natural history is the practice of falling in love with the world.

Thomas Lowe Fleischner

I first fell in love with the world of trees when I was a child. Most afternoons after school, I would grab a snack and a book, and scramble up one of the eight Sugar Maple trees that lined the driveway of my parents' home in suburban Maryland. Each tree offered its own vertical pathway to a comfortable nest aloft, a portal to places apart from the hectic world of my four siblings, four pets, parental directives, homework, and the ground-bound humdrum of the everyday. I could check on the progress of squirrel nest construction and feel the strong, trustworthy limbs of those trees holding me up for as long as I wished. The first book I authored was a sheaf of stapled notebook pages titled *Be Among the Birds*, an illustrated guide to tree-climbing, with an intended audience of fellow eight-year-olds, only one copy ever printed.

On one of those afternoons, I took a private oath that I would become a grownup who would protect trees. I imagined becoming a forest ranger or some sort of tree doctor. But in college, I discovered the world of forest ecology through the lectures of a natural historian, Professor Jon Waage, who studied the behavior of damselflies. His work as a grownup was to sit at stream edges, observing and recording their movements, getting answers to his questions: How does a female's wing size affect mate choice? How does mate choice affect population diversity? How does diversity affect resilience to community disturbance from human activities? These seemingly narrow questions ultimately related to much broader issues about life and death, competition and mutualism, and the evolution of life on Earth.

So I fell in love with the natural world a second time, through the study of science. The word *science* comes from the Latin *scire*, which means "to know to the fullest extent possible." Its practitioners use observation, experimentation, and quantitative documentation to meet the challenge of untangling the endless puzzles of nature. I could not resist its promise to deepen my understanding and connection to trees, and so I joined the world of academia.

During my first summer as a doctoral student, I took a field course in tropical biology in Costa Rica. Whenever our group of students and faculty set out on a rainforest trail, my eyes went upward to the plants and animals that I saw in the treetops. One hundred questions burbled up. How did the plants live up there without connections to the soil? What exactly were the birds feeding on? Who was pollinating which flowers? Were there insects that lived their entire lives up there? My instructors had no answers for these questions, because almost no one had studied—or even climbed into—the forest canopy. Most of these tropical trees have unnervingly long straight trunks with no branches for one hundred feet, rendering childhood tree-climbing skills useless.

By luck, I encountered Don Perry, a graduate student who was studying pollination systems in the canopy. He had applied mountain-climbing techniques to gain safe and non-destructive access into the treetops. As I watched him climb swiftly and safely from the ground to a small wooden platform 125 feet above, I gave a little scream of joy to realize that I had found a way to merge the two entities I most loved—tree-climbing and science. He taught me his techniques, and that began my graduate studies and my career.

As part of my first job as a professor, I returned to Costa Rica to establish a study site in the pristine tropical cloud forest of the Monteverde Cloud Forest Reserve. When I perched in the crowns of those huge trees, I was nearly overwhelmed by the sheer abundance and complexity of canopy-dwelling plants—a panoply of orchids, bromeliads, ferns, mosses, and lichens. These plants, called "epiphytes," have no connection to the vascular system of the tree (the tubes and chambers responsible for transporting water, sugars, and minerals from one tree part to another). Rather, epiphytes derive their water and nutrients from minerals dissolved in rainfall and intercepted by the leaves. Underneath these communities of living plants lies a fascinating layer of arboreal soil, comprised of the dead and decomposing epiphytes, inhabited by treetop versions of traditionally terrestrial invertebrates—beetles, ants, springtails, bacteria, and even earthworms, living out their entire life cycle high above the forest floor. The central question they evoked in me was how these arboreal communities might function as independent but interacting subsystems of the whole forest. From the canopy, I could see that my perspective of the forest from the forest floor was terribly limited; the forest had been a set of static and isolated trunks imbedded in the soil. Aloft, I perceived the forest as a dynamic tapestry of interacting plants, animals, and microbes, each contributing to the forest fabric: complex, connected, useful, strong, fragile, and beautiful.

Over the last three decades, my colleagues, students, and I have sought to answer that question. Canopy researchers have developed other techniques to study the arboreal world: hot-air balloons, treetop walkways, hanging platforms, and thirty-story construction cranes. We collected samples of epiphytes and analyzed their nutrient content to calculate their mass in the context of the whole ecosystem. We discovered that some trees put out "canopy roots" from their own branches and trunks, which gain access to the arboreal soil that accumulates beneath mats of epiphytes. We measured the amounts of nitrogen and other nutrients that the epiphytes intercept and retain from rain, mist, and dust, which can be as high as sixty percent of the total input from those sources. A summer-long study that involved perching on platforms in trees for six hours each day showed us that birds of the cloud forest use epiphytic flowers and fruits for over one-third of all of their foraging visits, demonstrating the importance of these plants to arboreal animals.

The small but fervent community of natural historians who investigate the canopy has documented the importance of canopy organisms at other sites around the world. Canopy biota produce large amounts of oxygen tree canopies, store carbon dioxide, protect soil, retain water, and support wildlife species. Urban foresters have documented the "ecosystem services" provided by trees in urban settings: reduction in noise, temperature, and pollutants. Thus, the growing body of treetop research documents that presence and protection of canopy diversity and function is essential to landscapes and humans who live in them.

However, human practices have drastically reduced

tropical and temperate forest cover with a concomitant loss of the functions that canopy-dwelling biota provide. Although immense amounts of knowledge are contained in the science libraries around the world, rates of deforestation, climate change, species invasion, and over-consumption of tree-derived products are increasing. Humans—especially those living in urban environments and working in windowless cubicles—are more and more separated from their connections to trees, soils, and wind. Midway into my academic career, I realized that communication of all of my scientific findings to scientists—through my academic papers and talks at ecology conferences—did very little to fulfill my childhood dream of being a grownup who protected trees. How might I fulfill that commitment of protection I made atop those Sugar Maples long ago? To foster a greater sense of mindfulness of trees, I needed an approach that goes beyond the facts and figures of science.

I knew that the simple act of climbing from the forest floor to the top of a tree changes my perspective of the forest. So I reasoned that mindfulness might emerge by borrowing and lending perspectives and modes of communication from other ways of knowing the world. This approach came to me when I climbed Mt. Rainier the previous summer. As I struggled up a snow-blanketed ridge of the tallest mountain in the Pacific Northwest, I diverted my mind from my aching legs by categorizing and giving my own names to the different types of snow I slogged through: snow that comes in humps, snow that comes in lumps, snow that sticks to my darn crampons. Of course, people familiar with snow have already created names for different types of snow, and I wondered if my snow taxonomy would match theirs. Perhaps my status as a "snow novice" would provide new insights into snow. That led to the idea that "forest novices"—people who never used the scientific approach to understand them—and people who have never even seen trees—might weave in valuable insights into the nature of trees and forests and provide a different and more heightened mindfulness of them. I envisioned a tapestry of interwoven threads of understanding, each thread representing the deep knowledge of a particular discipline or way of knowing, collectively creating a better understanding of the natural world. Like a real tapestry, it would create something complex, connected, useful, strong, fragile, and beautiful.

So, in 2003, I began staging a set of "Canopy Confluences," week-long gatherings in remote forest sites to which I invited artists, writers, poets, dancers, rappers, musicians, loggers, and tree novices including tundra-dwelling Inuits and people who are blind. The simple construct of these Confluences was for them to join me and my students in the treetops so that we might together explore the forest canopy and articulate what we perceive, with the hope of providing new insights to all participants and producing materials that might raise awareness of the values of trees and forests, instilling in ourselves and others a broader and deeper sense of stewardship, a way that grownups could better protect the arboreal world.

Our first staging area was Ellsworth Creek, which protects primary and secondary forests of southwestern Washington. My students and I rigged trees with climbing ropes, and hoisted up four plywood platforms for participants to sit upon. On the ground, we set up tents, a cooking shelter, and a campfire pit for two weeks of living in the forest. In the course of these interactions, participants taught me a third way to love trees and natural history—by providing me with the lenses through which they see the forest and express what they perceive. Two of the participants, a modern dancer and an

Inuit, gave me the greatest insights.

Jodi Lomask is the artistic director and choreographer for a San Francisco-based professional modern dance troupe called Capacitor, Inc. She wished to create a dance about symbiosis in tropical rainforests, and wanted to incorporate canopy interactions in her dance. I invited her to join that first Canopy Confluence, to provide her with scientifically sound information about canopy ecological interactions and to learn from her how dance informs and communicates "forest" to those who are attuned to movement and music.

After that first taste of the canopy biota of the Pacific Northwest, she and her seven dancers travelled to my study sites in Monteverde. We spent a marvelous ten days exploring, explaining, and interpreting the flora and fauna through the medium of dance. I taught them how to climb one of my study trees, named Figuerola, a huge Strangler Fig. The limbs of this cloud forest giant stretch horizontally in all directions, and are covered with a riot of epiphytes of all shapes and sizes, from tiny mosses to gigantic arboreal shrubs. Aloft, the dancers took in all that they saw and smelled and felt, not with a measuring tape, clinometer, and Rite-in-the-Rain notebook, as my students and I would have done. Rather, they recorded what they experienced with gestures of their hands, curves of their arms, and arcs of their eyebrows. After several days in the cloud forest, they peeled off their clothes to merge more fully with the forest textures. I was filled with wonder as I watched human limbs parallel tree limbs and human trunks intertwining with twisted tree trunks.

Her troupe performed her piece—called "biome"—in Seattle and San Francisco to audiences who were drawn to dance rather than science. We invited local conservation groups to offer volunteer opportunities from tables in the lobby, and scores of audience members signed up after witnessing the performance. Through those events that interwove art and science, I realized the power that emerges when we combine different ways of knowing, different ways of practicing natural history.

I also wanted to learn how people who have never seen trees would perceive and value them, which might add to my own understanding of forests. Would they find trees frightening? Boring? Useful? Beautiful?

I invited an Inuit named Emil Arululak to our second Confluence. He traveled for four days from his tiny village of Arviat, in Nunavut, located at the same latitude as Nome, Alaska, to reach our site. Emil had never seen a tree before he arrived at our rainforest study site the day before, and he had rarely even climbed stairs higher than the single two-story building in Arviat. By merely being in the forest, he was exploring new territory for himself. By climbing into the canopy, he was entering a whole new world. From Emil, I learned that the native language of Inuktitut had, as expected, over thirty-five words for snow, which included all of the snow types I had encountered on my ascent of Mt. Rainier. However, Emil had no word at all for "tree." "We use the word *nabaaqtut*," Emil explained, "which means 'pole.'" And forest? "We use *nabaaqtut juit*, which means 'many poles.'" Emil had no linguistic connection to trees at all.

At the tree-climbing site, Emil took his first slow climb up a hemlock tree. On a parallel rope, I accompanied him to the platform hanging sixty feet above the forest floor. Emil pulled himself over the handrail, and settled himself into a corner of the platform. He watched the swaying curtain of foliage that enveloped the two of us. He observed the sun reflected in the drops of dew on the underside of the spruce needles, and pointed out the pattern of sword fern leaves arrayed, fan-like, below us. After an hour, he nodded twice, and then asked to descend. In the days that followed in our little camp, Emil never climbed again, but I would see him walk up to certain trees—most frequently, western red cedars, the most important species to the Native Americans of the Pacific Northwest coast—and place his hands on them for minutes at a time. He spoke with each of the project participants about the forest and took notes in the hieroglyphic-looking script of Inuktitut in his notebook.

At the end of our time in the forest, we lit a last celebratory campfire. Each participant shared a song, a plant collection, a drawing, a poem, or a scientific question. At his turn, Emil slowly stood to his full height of just over five feet. He looked at each of us and said, "In these days, I have learned that trees are more than just poles. You must treat these big trees the way we treat the elders in our village—with great care and great respect. Trees are as important to you as our grandparents are to us because they teach you things." A long pause followed Emil's words. The strength of his insight came from its source—an individual who, until two weeks ago, had no personal, cultural, or linguistic connection to trees, but who was able to perceive and articulate the universality of values of trees for humans.

During these multifaceted immersions, I found myself falling in love with the forest all over again. The different insights and ways of articulating their understanding provided for me a new affinity between trees and people. The word affinity, from the Latin word *affinis*, indicates a relation by marriage. Humans and trees hail from different families, but the connections that Jodi and Emil provided made me consider myself as being married into a family of trees, with the challenges, responsibilities, and benefits that come with being so linked.

Perhaps other such partnerships between scientists and others will create ways of increasing mindfulness about nature.

When one is in love—especially with something as huge and beautiful and complex as trees—there is an urge to share this emotion with everyone, especially to those who have no opportunity to experience such feelings themselves. As my love of trees and canopy biota expanded, I sought to share my connections to nature with people who live in places where it is absent, just as a new bride might urge those sitting on the sideline of her wedding party to find a dancing partner. It occurred to me that the people who live in venues that epitomize the most severe endpoint of environments without nature are those who are incarcerated in prisons and jails, the spaces where nature is not.

In 2003, I started a research project that brought together plants and prisoners. I realized that it would be unrealistic to bring trees to inmates, but I could bring canopy-dwelling mosses inside the concrete walls to connect convicts with living, growing things that need their care. This "Moss-in-Prisons" project included prisoners in a combination research/ conservation effort to counteract the destructive effects of collecting wild-grown moss from old-growth forests for the floral trade. Florists, who use moss for their flower arrangements and to pack bulbs for shipment, have created a growing market for mosses harvested from old-growth forests in the Pacific Northwest. Since 2005, the moss-harvesting industry reached an economic value of nearly $260 million each year.

Ecologists have raised concerns about this expansion of this "'secondary forest product'" because they have documented that these moss communities fill important ecosystem roles. They take over three decades to regenerate, far longer than what would make for sustainable harvest at present removal rates from these ancient forests. No protocols exist for growing mosses commercially, or in large quantities. If I could learn how best to grow commercially usable moss, perhaps I could create a more sustainable source of moss and relieve the pressure of wild-collecting from old-growth forests. To do so, I needed help from people who have long periods of time available to observe and measure the growing mosses, access to extensive space; and, most important, fresh eyes and minds to put forward innovative solutions. These qualities, I thought, might be shared by many people in prison.

The biology of mosses also makes them suitable for novice botanists, because mosses possess "poikilohydric" foliage, which means their thin foliage wet and dry rapidly, allowing them to survive drying without damage and to resume growth quickly after re-wetting. Some mosses that have lain in herbarium drawers for over one hundred years have been revived by simply applying a little water and bringing them into the light, reawakened after a century of dormancy in the dark. They therefore tend to be resilient, a characteristic that increased the probability that the prisoners would succeed in nurturing living things.

After scouting prisons in my region, I found the Cedar Creek Correctional Center in Littlerock, Washington, directed by Superintendent Dan Pacholke, open to the program. From the beginning, he facilitated all aspects of the project, forging pathways through the Department of Corrections administration. We wished to know which species grow the fastest, and the inmates learned how to distinguish the different types of mosses, built a small greenhouse with recycled lumber, and took notes with the notebooks and pencils I distributed. After eighteen months, we all shared the excitement of knowing which mosses grew fastest.

There were other rewards that I had not foreseen, small and individual, but real. One of the prisoners, Inmate Hunter, joined the horticulture program at the local community college after his release, with a career goal of opening his own plant nursery. "I don't want to just mow lawns and trim hedges anymore," he said firmly. "I want to grow real plants." Another, Inmate Juarez, told me he had taken an extra mesh bag of moss from the greenhouse and placed it inside the drawer of his bedside night table. Each morning, he told me, he opened the drawer to see if the moss was still alive. "And though it's been shut up in a dark place for so long, it's still alive and growing this morning," he said, grinning. And then, more quietly, "Like me."

This "Moss-in-Prisons" project answered the scientific question I posed, which I valued from the standpoint of a researcher. However, the activities also resulted in better social interactions among the inmates, which was viewed positively by the administrators. The work also provided stimulation and a strong sense of contributing to the Earth, which proved to be of value for the inmates themselves. The superintendent requested other projects, so we brought in faculty to provide science lectures and initiate other conservation projects. These included captive rearing of the endangered Oregon Spotted Frog, the Taylor Checkerspot Butterfly, and seventeen species of rare prairie plants for ecological restoration projects around the state. The practice of inviting incarcerated men and women to actively participate in conservation has now spread across the country to many state prisons and county jails.

Although I felt strong satisfaction in sharing the love of practicing natural history with the inmates we were able to reach in the minimum- and medium-security portions of these prisons, I also felt compelled to find ways to bring nature to those in the deepest reaches of the prison system—men and women in the cellblocks of solitary confinement, where they are held in concrete windowless cells the size of a parking space for twenty-three hours a day, with one hour in a slightly larger concrete exercise room. We could not bring endangered animals and plants—or even lecturers—to these locales because of the high security protocols.

The human environment of hospitals is in many ways similar to those of prisons. The "inmates" of both prisons and hospital wards experience extreme stress and anxiety, as their activities and fate are no longer under their own control. Interior spaces are stark and sterile—for punitive and security reasons for prisoners; for health reasons for patients. Their webs of social interactions are entirely dependent on who might choose to visit them; often these individuals are islands in a frightening sea. Behavioral psychologists have documented that the view of nature outside a window or portrayed on backlit panels can reduce stress and speed recovery. In 2013, I found a maximum security prison in Oregon that was open to the idea of showing nature videos to men in their solitary confinement cellblocks to explore whether this might reduce agitation, anxiety, and the violent infractions that cause injury to inmates and officers. We installed a projector in the exercise room of one of the cellblocks and provided inmates with the opportunity to view the videos during their exercise time—one hour a day, three days a week.

After a year, our surveys and interviews of staff and inmates revealed that they felt lower stress, agitation, and irritability, and were able to carry a "sense of calmness" from seeing the nature video when they returned to their individual cells. Most significantly, we learned that the inmates who viewed nature videos committed twenty-six percent fewer violent infractions than those who did not view them, a convincing result for the prison officers and administrators—and for ourselves. Further work is now needed to learn how this "nature intervention" might work in other prisons, and to understand which elements of nature were most effective in bringing light to the darkest parts of our prison system.

I have been intimate with trees—through the curious eyes of a tree-climbing child, the number-filled notebooks of an academic scientist, the borrowed lenses from people of diverse disciplines and experiences, and most importantly, moving the shuttle of a loom that brings together the intersecting threads of nature and the multiple ways that society comes to perceive and communicate insights about our world. Practicing natural history—and the love that grows organically from that action—is a critical thread in the tapestry that makes up our world, an entity that is complex, connected, useful, strong, fragile, and beautiful.

Plants, Health, and People of the Forbidden Mountains

ALBERTO BÚRQUEZ

The study of natural history is the cleanest link between science and the use and enjoyment of the environment. An understanding of natural history brings the realization that we humans are an intrinsic part of the fabric of nature. The superb naturalist Edward O. Wilson framed the idea of knowledge and value of nature in a nutshell: "to the degree that we come to understand other organisms, we will place a greater value on them, and on ourselves." This assertion is especially true when encountering environments near pristine conditions, environments that have the quality of being beautiful and sublime, locations that are, for many people, sacred spaces. Places like the forests of Yucatán explored by Stephens and Catherwood, or John Muir's Sierra Nevada offer two examples of the mixture of the sublime unknown: terrifying and beautiful. Powerful landscapes so rich in biological diversity that to the initiated they offer all sorts of resources. Spaces hosting a multitude of molecules and compounds created by living organisms—food and medicines—to feed and heal the body. Also, these are domains where love and nature are embodied into a single entity fusing the physical reality and the metaphysical world of the perceptual mind (aside from any discussion between dualism and physicalism). The intensity with which these qualities are observed and understood is likely to correlate with the nourishment we have provided to the naturalist we all have inside.

The Naturalist Within

In our times, few, if any, places on Earth can be classified as pristine, but many still have unique qualities related to conservation and naturalness. Some are so outstanding that society as a whole agrees on protecting them as national parks, nature reserves, and world heritage sites. Even if the natural spaces are degraded and spoiled, it is easy to find beauty in the ways that nature reassembles itself, whether it is a derelict industrial park, or land after a recent volcanic eruption. Both sites offer opportunities for colonization by organisms. Both are avenues for the mystery of life to be revealed.

When studying island biogeography, I was moved by the image of a lone spider sitting in an empty web on the newly emerged volcanic island of Anak Krakatau. The island, arising from the sea in August 1930, was completely sterile new land created by the combined effects of lava flows, nuées ardentes, and tsunamis. The whole biological diversity of the microcosm in formation was represented by that one spider. As a scientist, I knew that the newcomer was doomed because it was completely alone. It had arrived before plants, organisms that have the ability to survive only with sunshine and a little rain, providing food and shelter to other species. The lone spider is the first element in the slow accretion of plants and animals that will eventually lead to a complex jungle—Darwin's entangled bank—rich in biological diversity and complex interactions. For the naturalist, as well as for the artist, the simplicity of a small spider in an expansive field of dark volcanic cinders made an outstanding composition that elicited deep feelings. From the point of view of the spider, however, biodiversity was the key to survival.

Owning a Farm

In my trade—research ecologist and conservation biologist—the appreciation of nature and its conservation arises from two main considerations: one purely practical, related to the usefulness of keeping the proper functioning of ecosystems to provide goods and services for human well-being, and another from a philosophical and ethical standpoint, related to the way in which we appropriate natural resources and the need to restore the environment after disturbances. To Aldo Leopold, an iconic figure in the development of modern environmental ethics and the discipline of wilderness conservation, land was, until recently, treated as a commodity instead of a community to which we are tied, belong, and depend. Leopold clearly described the ever increasing rift between nature and natural resource use: "There are two spiritual dangers in not owning a farm. One is the danger of supposing that breakfast comes from the grocery, and the other that heat comes from the furnace." Leopold left implicit the idea that living on a farm provided an intimate knowledge of nature. That knowledge, in some cases, transcends the utilitarian value of farming and paves the path for the budding naturalist. Today, only naturalists and shamans—powerful naturalists from our deep past—could be aware of the dimension of nature's values. My friend Michael Soulé, a key player in modern conservation biology, clearly stated that "biotic diversity has intrinsic value, irrespective of its instrumental or utilitarian value." Years later, scientists started ascribing value to the priceless services freely provided by nature, at least to convince recalcitrant economists and politicians that protecting the environment was in many cases a wise monetary investment.

Resources, Ecosystem Services, and the Lightness of Being

People longing to leave their homeland are unhappy persons, said Czech novelist Milan Kundera. Many people live oblivious of the marvels surrounding them, some others long to leave. But a few become natural historians, either by formal or informal inheritance of their elders' knowledge, or as *de novo* natural historians. These usually stay or return to their childhood place. The knowledge of the natural world represents a survival tool indisputably shaped by evolution. This knowledge is not, however, a gift to preserve environments in their original pristine condition, but more usually to transform and extract useful resources. Major human revolutions attest to this: from prehistoric changes like the Clovis culture that accelerated the demise of the Pleistocene megafauna, changing forever a pristine American landscape; to the Neolithic revolution in Europe that removed the extensive European forests; to the present calamities of global change and biodiversity loss. Although it is likely that once and again the concern for preserving natural resources arose in ancient societies, mainly as a response to vital resource depletion, it was not until much later that human concerns for conservation started. To some, John Evelyn's work *Sylva*, written in 1662, represents for the first time the formalization of these concerns. The idea of assessing and valuing ecosystem services is a direct descendent of these practical efforts to preserve natural resources—in the case of Evelyn, to plant trees to provide timber for His Majesty's ships. On a broad scale, ecosystems provide direct services related to the support of human life like food, materials for shelter, and fuel, and also a myriad of indirect services such as clean water and air, detoxification of wastes, pollination of crops, and maintenance of soil fertility. Ecosystems also provide intangible services like well-being, comfort, and joy. Perhaps the unhappiness leading to escape from the homeland is a consequence of utter ignorance on the role of natural resources in our lives. Is perchance the life of the Inuit less rich than that of Amazonian people? Surely not!

The Elusiveness of Well-being and Mental Health

A great effort has been devoted to the appraisal of the many services ecosystems provided to humankind, particularly those that directly represent an economic benefit or those that are perceived as major threats. Many researchers—Francesca Grifo, Joshua Rosenthal, Eric Chivian, and many others—have highlighted the direct impact of biodiversity on human health. Most of their contributions related to the active role of nature in providing medicinal plants and their chemical compounds for therapeutic uses. More difficult has been the assessment of more elusive values like well-being, comfort, nature enjoyment, spiritual inspiration, education, and mental health. The publication of Roger Ulrich's research in 1984 promised revolutionary healing therapies involving exposure to "greenness" and biodiversity. He showed that hospital patients whose rooms were overlooking trees recovered more rapidly after surgery, required less medication, and had fewer complications than patients in wall-view rooms. His results have been replicated in urban environments by Richard Fuller and colleagues.

However, the demonstration of a link between health, well-being, or improved survival and living in a better-quality natural environment has been more elusive, especially at large scales when, for example, Dutch scientist M.M. Huynen and associates found no relationship between health and loss of biodiversity on a global scale. A study that I co-authored demonstrated a significant but very tenuous relationship between biodiversity and depression here in México. The multifactorial nature of well-being—so subjective in perception and so difficult to assess in field conditions—makes it challenging to demonstrate the link outside the hospital room or the urban landscape. The association between well-being and natural places that many people from urban regions clearly recognize, like the recent finding that urbanites that walk in nature are less prone to depression, becomes muddled when studying people living in rural environments where variables such as income, exposure to pesticides, toxics from mining, smoke from forest fires, and other hazards from the rural milieu are considered. These hazards also include those arising from biodiversity itself, such as emergent diseases like Ebola or HIV, or the myriad of blood-sucking and poisonous organisms that populate biodiversity hotspots. Even though the Galilean motto, *e pur si muove*—and yet it moves—applies when going to sacred forests, to pristine lakes, or to mountain summits, what is lost to exposure is gained in well-being.

Being a field ecologist brings the great privilege of working in the outdoors. During strenuous workdays there is always the opportunity to explore and enjoy outstanding wild places with splendid vistas and unique biological richness. Of course, I do not have to plough the land or take care of cattle, but carry out my work of measuring and implementing natural and planned experiments. During an autumn field trip to the Sierra Madre, my dear friend and mentor Paul S. Martin, creator of the concept of rewilding America with mammoths and other Pleistocene megafauna, remarked to me that the tropical dry forests of western México had a gaudy autumn color display—in some places rivaling the forests of New England. The remoteness of these forests and the brevity of the display had sheltered this secret, and the hard life of its inhabitants, as with the pioneer New Englanders, provided little time to ponder about nature's autumn pageant.

In these dry forests, as in any other remote areas of the world, biodiversity and human health are closely intertwined because the food comes from the land—beans, squashes, corn, and a variety of forest products—and health depends on the knowledge of the vast natural pharmacopeia of healers and shamans. One such place lies in the high Sierra Madre Mountains of southern Sonora, México, where the Guarijío indigenous people live. Close contact with these gracious local people and their land sparked in me an understanding of the concept of sublime.

A Trip to the Forbidden Mountains

Under the dark, fragrant canopy of dryland subtropical trees—*mautos*, *torotes*, and *josos*—up to sixty feet tall, the path wanders, and our feet break the red and yellow carpet of dead leaves. Kapok trees guard with thick spines, their pods dangling from high branches. These pods are full of silken strands used for soft pillows. Against the shimmering blue above, the *Bursera* trees, in their last breath of summertime, write goodbye to their scarlet leaves on golden scrolls of discarded papery bark. My friends Cipriano Buitimea and David Yetman climb the steep slope with apparent ease.

The road from Álamos to Guajaray, until recently an endless domain of tropical deciduous forests—*monte mojino*—is now peppered with clearings—some small and some covering whole mountainsides. This sloppy mosaic shows the forest as a patchwork of rich and poor fabric fragments. Some patches—the rich ones—harbor a treasure-trove of medicinal, useful and edible plants that we still don't know how to exploit, but that the Guarijío have used for centuries. The others resemble depauperate African savannas oddly transplanted to the American tropics because they are dominated by the invasive African Buffelgrass planted by cattle growers for their herds.

We have entered the land of the Guarijíos, people that the Mexicans say disappear like ghosts in the forest. We are among high sierras of vibrant names: *Sierra Charuco, Sierra Saguaribo, Sierra Cajurichi, Sierra Chuchupate, Sierra Canelo.* Forbidden mountains.

David and I talk with Cipriano about the traditional uses of plants—some with no specific recipes, others including complex preparations and mixtures. Remedies for common ailments like bruises and cataracts and remedies for mind disorders caused by misfortune, curses, and spells. He speaks softly, modestly, wisely. Cipriano is the last of a long line of traditional healers and has a deep understanding of nature and natural history. I can feel the deep empathy emanating from this shaman. Just by listening, sitting on a fallen log of the humid forest of early autumn, I can find peace. I recognize that this insightful man is able to cure all sorts of illnesses because of his intimate knowledge of the forest based on the collective knowledge passed along countless generations.

Trying to cut a tough botanical specimen with a dull knife, I end up with a deep wound in my hand. Cipriano notices it and asks me to clean it with water from the nearby creek while he rummages around. He comes back with several soft and pliable cocoons, neatly cuts them in strips, and dresses my injury. The next day I am told that these were black widow spider cocoons; by then the wound has almost miraculously healed.

Much later, I discover that an important line of research in medicine, biotechnology, and polymer science is on the value of spider silk to heal wounds.

Reaching the *mahuechi*, the native name for the small plots where vegetation is slashed and burned to cultivate the land, we move among corn plants, some ears showing in the split husk white and purple smiles. We sort the large pumpkins that start changing their dress from a formal green to a gay orange. Because of their small size, rarely over two hundred yards long, and the short time of use, the *mahuechis* allow the regeneration of the forest. Nowadays, below the mountains and toward the cities, the ancestral slash-and-burn subsistence agriculture is being replaced by bulldozer clearings to establish African forage for cattle. Extensive clearings cover thousands of acres where exotic grasses and fire will hardly ever allow the return of the native vegetation and natural forest dynamics.

Cipriano's house sits atop a small red hill. Here, his daughters are preparing tortillas from their own locally harvested purplish corn. There is the clay pot where the corn is cooked and left to soak in a highly alkaline lime solution until the corn seeds become tender, lose their tough skin, release some essential amino acids, and the starch grains dissolve and jellify. Close to the pot is the *metate*, a large stone grinder where the cooked corn grains—the *nixtamal*—is milled to form the dough—the *masa*. The tortillas, after being rolled and flipped into the girl's palms, are gently positioned on the *comal*, a smooth, circular convex griddle or clay pan sitting atop the cooking fireplace. While cooking, chortling and giggling, they carefully turn and press the tortillas against the *comal* with a clean towel. This traditional process creates moist, fragrant, utterly delicious tortillas—the best I have ever had in my life!

Amidst an Indian summer, we start our way back. After the summer monsoon excesses, the river—Rio Mayo—is no longer a wild roaring beast, and its waters are shallow enough to wade. A few youngsters splash in the quiet pools. We stop in the town square of San Bernardo near the house of Howard Scott Gentry, a prominent naturalist of the 1930s. He rambled all over the Mayo basin making botanical collections in an effort that, even today, with roads and bridges, seems almost impossible given the oppressive summer heat, the wild rivers, and the precipitous terrain.

Here, the first important scientific work on the American tropical deciduous forests was conducted. Gentry described the ecological structure of the forest, and the diversity of species, environments, and people, opening to the world the gates of the Guarijío paradise.

We forded Arroyo Taymuco to reach *la otra banda*—the other side of the river—the Guarijío neighborhood of San Bernardo. Up there, at a village named *Los Jacales* ("The Huts") after climbing a steep hill covered with *Lignum vitae* trees, we sit quietly, protected by the cool *ramada*, awaiting the musicians' arrival: a violin and a guitar. The music, repeating itself once and again, has subtle variations that echo the richness of the *monte*: a mosaic of slightly different forests, where each site offers different species and resources. A barefoot teenager, a would-be *pascola* dancer, hops to the cadence of the *son*—Mexican folk music that merges indigenous and European forms—played by the wrinkled, venerable, skillful indigenous virtuosos. The European violin and guitar sharply contrast with the Indian *pascola* dancing with the resonant *tenevoim*—sand-filled cocoons of a giant saturniid moth that are worn in long strips like cartridge belts wrapped around the dancers' calves. The absence of the traditional cane flute and hide drum bespeak the slow loss of traditions and knowledge.

For a magical second, the solar gold loses its edge and the warm air freezes. For an instant, the music embraces the land, the forest, and the people. Far away, the crow cries and, with nostalgia, we realize that the gates of paradise are closing. Our autumn rendezvous with the forbidden mountains has almost finished.

Sunday School

GWEN ANNETTE HEISTAND

I was three when I first fell in love with light. Quivering with the sweet scent of hyacinths and morning dew pooled in daffodils' yellow throats, I spun in circles, humming and buzzing, trying to touch all the points of reflected light in my field of vision. The pleats on my lavender pinafore opened out, my body perfectly centered in a hoop of fringing white eyelet lace. Each blade of grass, every drop of water shed tiny suns into the vortex that was me. Squinting my eyes and whirling faster and faster I could change light into streaks and smears, paint with it, collect it, swallow it, and sing it back out. I was sun and sound and space. I smelled like Easter, like spring, like damp living soil. The world was screaming to be noticed and I was smack dab in the middle answering it note for note. My mother, trying to hold a conversation with our neighbor, asked me to settle down. How could she not stop everything and spin and laugh and shriek with delight?

When I was nine we moved to acreage in New York State that contained broad-leafed hardwoods, headwaters of the Mianus River, and was connected to vast tracts of open space. I was miserable in the world of fifth grade and school bus politics. There was a place by our brook where sunlight through green leaves was so beautiful, the smell of damp, black earth so heady, fairy rings so well-appointed, stream chortles and gurgles so enticing, brush of wind on my cheeks so intimate, that I would slip off my clothes and lower my body into the dark silty ooze. Cool mud spread down my backbone and into the hollows behind my knees. I needed every pore of me, every cell, open and receptive. I wanted to ingest colors and invert myself into the world. I could feel my long, curly hair merging with roots and tricklings. My body became a surface to absorb sunlight and nutrients. Two eyes, a nose, and two ears poking out of leaf litter; the rest of me tree and wind and light and earth. Murmur of brook; fly buzz; fish splash; towhee's *drink-your-tea*; my beating heart; one leaf brushing another; a beetle's gentle steps by my ear.

Thirty years later, I'm sitting in large interior windowless rooms of an office high-rise. The fan in the overhead projector drones on, fluorescent lights hum and flicker. Machines whirr. It is a world of recirculated air, artificial colors, the glow of computer screens, and "inspirational" posters with photos of eagles and sunrises. In every meeting, I amuse myself by asking who I could imagine sleeping with—the answer always a resounding *no one here*. This gives me pause. I wonder if the human race is speciating on ethical lines. I am often the only woman in the room. The men joke about gangbangs and skid marks on their underwear. There is nothing remotely like life bouncing off of me. No sound or color or light for me to absorb and reflect. Torpor sets in. I am contracted and tight—a dry, waxy seed protecting myself. What fire or flood or act of god will place me in fertile ground so I can open once more?

Open where? What was I suited for after a work life spent in software design, banking, and transportation? I was told I had an enviable skill set, a marketable resume. I helped teams tackling container storage and detention charges at rail yards, crane operations at ports, perceived discrimination in work groups, and proper shipment of hazardous materials. I taught statistics to longshoremen, process management to truckers, database design to bankers, and facilitation to executives. I met with women salespeople who felt marginalized because their male coworkers conducted business in strip clubs and expensed prostitutes in Asia. I developed strategic five-year plans and metrics that linked financials to business processes. Memphis, Singapore, Detroit, Chicago, Long Beach, Seattle, Atlanta, Oakland. I told myself I was helping to effect change from within. And yet I felt an uneasy complicity, shame, and despair. When I listed my accomplishments, tried to craft a CV, I couldn't come up with anything that was meaningful to me. Woven through everything was fear. Fear that I'd be destitute, unemployable, and unable to support myself if I left this lucrative world. I wanted to wake up from what felt like an enchanted sleep. To remember.

I once watched a Great Blue Heron hunt for her dinner at the edge of the Sinkyone wilderness in Northern California. Her stillness charged with electricity and singleness of purpose, taut with potential, yet relaxed and infused with experience and instinct. Her strike hallucinatory in its speed—cancer crab dangling from her beak. Crab immediately released its captured appendage and dropped back into the pool. Nature's ingenuity providing crab with the potential for regenerating its limb and heron with a snack. When I think of my decades in the corporate world, I think of that crab. Shedding limb after limb—until what was left? Corporate. Corporal. Corpus. Body. How to re-member?

Limb regeneration in crabs happens as part of the molt

cycle. The old exoskeleton must be shed in order to grow, for limb buds to develop, to get rid of parasites, and to repair. My ecdysis started one warm Sunday morning in May. I rode my bicycle to Lake Anza in the Berkeley Hills determined to apply my skills as a systems analyst, facilitator, strategic planner, director, and instructor of process management to my own life. I swam, wrote, sunned, wrote, swam, made lists of questions, and defined my terms. Six hours and a modest sunburn later, my mission statement became one word: "joy." My choices would be informed by beauty and grace. I defined beauty as integrity of essence; grace as fluidity in change. My challenge, as I saw it, was that all my metrics, my references to space and time and currency, were of the world I wanted to leave. I hungered to know tide pools and how seasons change, which plants were my neighbors and the source of my water. I reasoned that if I focused on where I lived, the way I marked time would change from executive committee meetings and five year plans to nature's rhythms.

I began a ritual I dubbed *Sunday School*. Every Sunday I would explore these questions, let new ones emerge, make lists of what I wanted to learn and go and learn it. I would practice Sunday School for fifty-two weeks with an open mind and heart and see where it led. My only rule was that I spend one hour, at minimum, writing at the end of every Sunday.

૪ ૪ ૪

Sunday, June 11 (week 4)
My jeep is turning into a traveling library—natural history of all sorts; field guides to birds, plants, insects, spiders, shells, seaweeds; a growing collection of reference books and articles. I purchased a hand lens two weeks ago and some waterproof binoculars. Today, at McClure's Beach in the Point Reyes

headlands, armed with a second edition of Ed Ricketts' classic *Between Pacific Tides*, I learn that when young, *Hinnites giganteus*, the Purple-hinged Rock Scallop, swims around flapping its two valves like wings. Then, "the ways of placid old age creep on it rapidly. The half-buried undersurface of some great rock offers the appeal of a fireside nook to a sedentary scholar, and there the scallop settles." The lower shell distorting in order to cleave to the irregular surface of its rock, the upper shell adapting to fit. But how is the shell created?

It turns out, bit by bit. Slowly a lifetime is limned in calcium carbonate. Each tide. Each predatory attack. Each summer. Each winter. Recorded in checkmarks, ridges, grooves, repairs, and deposits. Signature drill holes from whelks, octopus, and moon snails hinting at the resident's demise. Shell morphology married to function and 500 million years of evolutionary interplay between hunter, hunted, and environment. All there to read for anyone who cares to learn the language.

If someone looked closely at my shell they might see thin layers from a young life spent moving: nine domiciles in as many years; a checkmark capturing the silence and unhappiness of fifth grade; wide, smooth layers from summers on our farm in Pennsylvania and sailing the eastern seaboard; ridges and bumps as I moved out into the world —college, New York City, Cambridge, Berkeley; signs of repair from emergency surgery at 26; decalcification from the acidic environment of my current tenure in Corporate America. How did I get from a childhood immersed in field and stream and marsh to an office building in downtown Oakland?

It turns out, bit by bit. Slowly a lifetime is limned by the decisions we make. Each relationship. Each move. Each crisis. Each job. I wonder where my shell records the moment when money and my fear of not having it become the currency of my life. Money is just a chit, an IOU. Currency in its pure form is the combination of energy and time required to live. From Latin *currentia*: a flowing. The same root as current: belonging to the present time; passing from one to another; a smooth and steady onward movement.

Sunday, June 25 (week 6)

Running in the hills this morning I heard a bird in flight and then another and then groups of birds. Time has slowed down and opened up enough for me to hear the muffled thwop of birds' wings. I'm gradually entering the world in which I want to live. There is a big hand on the knob changing the radio station. One world fading out, another coming in, albeit with a little static. One frequency contains car engines revving, elevator doors shutting, high heels clicking across marble lobby floors. It contains deadlines, timelines, killing time, TGIF, time is money. The other frequency contains birds' wings, fog drip, my feet hitting wet earth, and eucalyptus leaves chattering as a breeze catches them. It seems to me the two worlds don't belong together.

A spider, *Araneus diadematus*, has spun her web across my front door. I can watch her from my desk. I imagine suspending my web across the door frame under the porch light in prime spider real estate, my body extended by spider silk, sensing vibrations with lyre-shaped organs in my paired, jointed appendages. I watched her all through the wee hours this morning. How she put the finishing touches on her prey capture device, then employed it. Hanging head down in the center of her web for long periods, tarsae gently resting on radiating spokes, waiting. Vivaldi, Brahms, Faure, Grieg, Bach cycle through my CD player. I turn up the volume, shut my eyes, and put my hand on the wall and my foot on the door frame. I wonder if the spider is listening too, if she can tell the difference between Bach and Brahms. Maybe this is why I was so attuned to frequency on my run. I think about music composed in an era of computers and combustion engines and technology that allows us to wipe out dozens of species a day. As time slows down, my sense of urgency grows.

Wednesday, August 27 (week 15)
A few hours ago I excused myself from a five-year planning meeting to go to the ladies room and next thing I knew I was here, at North Beach. Apparently weekday bathroom breaks are now part of Sunday School. It is a glorious day. Crashing surf, salt spray, cloud patterns that look like crashing surf. There is no one on the beach with me except the usual suspects— pieces of jellyfish, crab carapaces, bull kelp. I walk. A rhythm is established between breath, feet, sand, expanse of water, dunes, sky. Step by step a conversation between my soul's landscape and the landscape's soul. I find my seat—a giant piece of drift-wood with a wraparound desk, just like in elementary school.

I love this edge where two worlds meet. Coastal strand. Stranded. Tide wrack. I belong. My home is on the edge, on that elusive line of wet and dry and shifting sand. And on fringing marsh and where meadow meets forest and alpine meets subalpine. The space between molecules, sighs between the words, smiles at the edge of consciousness. I have been operating in the shadow of an unspoken curfew. Promising to be back at a reasonable hour, to not spoil my party frock, to not talk to strangers, to settle down. Today I danced out the lobby doors and found myself in bare feet without money for a phone call, talking to strangers, skipping stones across the surface of my life.

In the ripples, time folds and brushes up against itself,

child and adult equally present. I travelled east this past fort-night for Grandma's funeral in Harrisburg, extending my trip to accompany Mom and Dad back to Florida. Cocooned in the car, we sang through Pennsylvania, Maryland, Virginia, North Carolina, South Carolina, Georgia. We talked of life and death, our heroes, our teachers, Grandma, places we'd lived, summers spent sailing, clamming, putting in fence. We re-membered, steeped in a different type of edge community where present and past, young and old, parent and child, east and west coalesce.

Who *were* my teachers, my heroes? Fields and forests and coastal strand certainly. Parents, grandparents, and god-parents that introduced me to sassafras tea and birdsong and the physics of wind and sail. Books where lessons were learned by becoming a pike or kestrel, where moors bloomed in patch-works of heather and gorse, where spirits of beeches and larch were made manifest in graceful dryads, where ancient ones whispered to old man willow, where children conversed with beavers and mice, where otters wreaked havoc in bathrooms, where bear and pig tracked themselves around a copse, where rat and mole shared the wonders of river life. They are my teachers again.

Enveloped in my driftwood desk, I surface. I meet myself in time.

I know what I have to do.

Sunday, October 22 (week 22)

This past Friday, I climbed out of darkness at the 12th Street BART station, through the ubiquitous human urine and vomit, into daylight, concrete, street people, business suits, and urban renewal. The green glass windows of my office building strut-ting up to the sky, reflecting light, and hiding what was inside.

It occurred to me that buildings are the sex parts, the fruiting bodies. Instead of root systems exchanging information and nutrients with fungal mycelia, tunnels filled with networked rails and human effluvia connecting to expressways and interstates. Instead of oak trees and mushrooms, skyscrapers and superstores. Instead of seeds and spores, ever-increasing numbers of products in containers and houses and landfill.

The night before, in my class on evolution of cetaceans at the California Academy of Sciences, I held the fossilized endocast of a 20-million-year-old dolphin brain. A brain that could echolocate and communicate when humans weren't even a gleam in nature's eye. *Twenty million years ago.*

I spent today picking up garbage on the coast. The juxtaposition of urban renewal and endocast and beach garbage got me wondering again about ethics and speciation and what draws one body to another. What are ethics but what we respect, what we cherish, what we nourish, what we protect? How *could* I be attracted to a person or a way of life that is party to destroying all I hold dear?

Sunday, November 26 (week 29)
My cousin and I went off to the Puget Sound this week to search for the elusive Giant Pacific Octopus and honor a pact we made over our grandmother's grave. I got my scuba certification in July and this was my first dive away from home surf. We descended into a world populated with hundreds of giant white-plumed anemones and every color of sponge, tunicate, and bryozoan covering an old ferry pier's pilings; fantastical nudibranchs writhing and dancing around us; purple, pink, orange, and gold giant sunflower stars coating the seafloor.

I read about sponges on the flight home. They have such great names: *Ophlitaspongia pennata, Halichondria panicea,*

Cliona celata, and my favorite just for the sheer joy of speaking it out loud, *Hymenamphiastra cyanocrypta*. Splotches of color that are colonies of single cells, specialized in their functions to work as a group and thus create an organism. Flagellated collar cells that capture food and digest it. Amoebocytes that engulf large particles and travel about in the mesenchyme bringing nutrients to the rest of the sponge. Cells that provide structure. Cells that reproduce. Cells whose sole purpose is to contain pores to allow water and food particles to enter the sponge. Here's the thing. Scientists have pushed sponges through a fine silk sieve to see what would happen. All those single cells started moving. After a couple of weeks individual cells found their way back to each other and became a sponge again.

I have pushed all my single selves through fine silk. I am watching the petri dish in anticipation as they re-form Gwen. Why did I need to separate? The homecoming, the reformation, is why. Understanding how my severed parts fit together. Learning how to take care of myself,—which parts engulf food, which are pore spaces, which reproduce, and what is the extracellular glue that holds me together?

On my way home, I drove directly from the airport to my goodbye party. I walked out of those glass lobby doors for the last time. No reference to space or time remaining. What had seemed like such a giant leap of faith—merely leaning into the door, pushing with my shoulder, and stepping across the threshold.

On my way home.

Sunday, September 8 (week 70)
I've been packing all morning. I start graduate school in three weeks—the new School of Environmental Science and Management at UC Santa Barbara. I feel like I'm leaving a

lover only my lover is a curve of coastline, an oak-dappled hill-side, mudflats sweating at low tide, scent of willows, thrush's flute, sun-released aroma of Bishop Pine. I want my leave-taking to be conscious, elegant, and generous. I want to take my time saying goodbye. I want everything I pack to be chosen; otherwise it will be recycled, repurposed, or gifted.

What to do with the odd curios that litter my shelves? A leather cigarette pouch branded with *Smoke and Cast Away All Cares* once belonging to a grandfather who died of lung cancer. From a grandmother who had a double mastectomy: *Your Bosom Friend for Folding Money*: "the pointed flap goes next to the body inside the brassiere and is held firmly in place. When traveling, your bosom friend will always protect your bills even if you lose your handbag." A Fuller powder brush "with imported hair" from a great aunt bald from chemotherapy. These things keep company with a dolphin's jaw and part of a wasps nest, a Burmese python's shed skin, heart cockles, a purple-hinged rock scallop, driftwood, sharks' teeth, a piece of dried sponge, horseshoe crab carapaces, seed pods, a fossil sea urchin, a perfectly preserved dragonfly removed from my car's front grill, coral tests, flowers pressed in an old notebook, a piece of birch bark, skate egg cases, a skull finally identified as brush rabbit, dried coyote melons, cholla skeletons, moose scat, hatched black snake eggs, a bouquet of northern flicker feathers, a whelk operculum. Flotsam and jetsam stranded as lives recede. *Mementos mori.*

I think about aggregating anemones, *Anthopleura elegantissima*, selectively covering themselves with bits of light-colored shells and stones as the tide withdraws. Not only do the fragments of shell and stone reduce exposed surface area, they also increase the anemones' albedo, keeping them cooler. These oddments scattered around my cottage kept me from drying out, reflected my life back to me. They are a map; Hansel's and Gretel's bread crumbs; winding through my months of Sunday School, my childhood, my life in the theater, my years in the corporate world.

Anthopleura elegantissima also harbors mutualistic algae in its tissues. Like good plants, the algal symbionts convert inorganic carbon into carbohydrates and release oxygen. The anemones provide their guests with concentrated carbon dioxide and a place to live where radiation required for photosynthesis is available. I find myself reflecting on the gifts of energy and time, the currency, I've exchanged on my journey. My teachers, human and more than human, have given me names of things and showed me how to find them. They have given me clear granitic tide pools; the delicate balance of salt marsh and why tides are; living tissue of bristlecone pines already old when western civilization began; corals feeding by moonlight off the Cayman Islands; booming sand in the Eureka Dunes; the way the first rains smell as they sweep across the Sonoran Desert. Sponge and scallop, dolphin and crab, bird and spider have shown me how to know a place and situate myself in the world. I understand on a cellular level the importance of ritual in creating relationship and why the metrics I choose to measure life matter. These lessons are incorporated in my tissues like *Anthopleura*'s algal symbionts. They bring with them a responsibility to use the gift of energy given.

❧ ❧ ❧

My year of Sunday School stretched into five years, ten years, twenty. A teacher of mine once stated that discipline is the dance partner of wildness. I know this for a fact, in my art, my work, and my life. I still spend every Sunday learning something. I'm resident biologist on a thousand-acre nature preserve that forms much of the eastern watershed of the Bolinas Lagoon, a misnamed estuary, formed over seven thousand years ago by tectonics. I sit in my porch chair on the North American Plate and look across the San Andreas Fault to the Pacific Plate slowly migrating up the coast. I teach people who teach children. We explore slime molds and spider silk and why birds sing. We marvel at pollination and decomposition and transpiration. Many days I open my office door to some gift with note attached—a putrid carcass of an opossum, "*Gwen, I found this by the driveway. Look there are dermestid beetles!*" The last molt of a dragonfly naiad, "*I saw the adult emerge! Sending pictures.*" Moist bobcat scat, "*positive verification . . . I watched it come out!*"

When this happens, I sometimes recall a different office in downtown Oakland, where bobcats and dragonflies and dermestid beetles were nowhere part of the lexicon. I think about complementarity in physics—about objects with properties that can't be measured and observed at the same time. Particle and wave. Mind and body. Corporation and human being. Natural world and world of commerce. The type of measurement determines which property becomes apparent. I am grateful that I understood this on some deep level that afternoon at Lake Anza. And I know healing requires opening to the paradox that both are present, finding the one within the other.

Flirting Dragonflies

BROOKE WILLIAMS

Between a dragonfly's two compound eyes, three simple eyes called the ocelli cluster on top of its head. These eyes measure light levels, orient the organism toward the horizon, and map visual space. In a 2006 paper, Australian scientists documented their discovery that the photoreceptors of the dragonfly's *median ocellus*—the middle of the three eyes—are connected via eleven long neurons down its neck directly to its motor centers. This helps explain why dragonflies are the most effective flying hunters in the animal kingdom.

The ocelli are represented by part number four of twenty that when properly assembled make up my Metal Earth Series "Dragonfly" facsimile. It is one of dozens of 3-D laser-cut models in the series, which includes a World War II bomber, a tank, a schooner, a windmill, and a train engine.

My friend Erin gave me the dragonfly model knowing about my fascination/obsession with these organisms. After four frustrating hours with needle-nosed pliers, a magnifying glass, and a pair of tweezers, bending pieces into proper form and putting nearly invisible tabs into slots so small they must have been made with a hypodermic needle, I held the finished dragonfly out in my hand, expecting it to fly.

Assuming that the size was somewhat accurate and noting the shape and that the hindwings are broader than the forewings, I surmised that I'd built a model of *Anax junius*— the Common Green Darner.

My model dragonfly matches the first vision I had a decade ago, which opened a door to a world more deep and intricate than any I'd previously imagined. This vision came in a dream I had while lying against a boulder on the edge of an island off the coast of British Columbia. In my dream, I found a small round stone into which a perfect dragonfly had been carved. An hour later, on rejoining my group, I noticed four small, bright blue dragonflies (I now know that they were damselflies, specifically dancers, of the genus *Argia*) bobbing among the bright-headed flowers in a wild garden. Time disappeared; I watched in awe as they hovered and darted after individual gnats swarming in a just-hatched cloud. When I pointed out these magnificent creatures to a colleague, she said, "There are hundreds of them. They've been here all week."

The poet Ezra Pound believed "the natural object is always the adequate symbol." When I first read this, I suspected he was referring to the mechanics of a poem. But with my dragonfly dream I discovered that attaching symbolic value to an actual natural object opens up a wide and wondrous world of possibility.

In all the biological field work I've done, and during the thousands of hours I've spent bird watching, I've seen many dragonflies. But I had not given them my quality attention before. Those blue dancers were actually *flirting* with me, that day, on that island, demanding my attention, absorbing all of it.

To Arnold Mindell, the founder of process-oriented psychology, *flirting* is when an archetype outside of us attracts our attention because our unconscious has important information for us. According to Carl Jung, an archetype is an image, thought pattern, or idea passed generation to generation in the same way that genes are passed on. Archetypes have meaning that is universally understood by all humans. Dragonflies have been on the earth for 300 million years (compared to our 200,000 year history) and must play a role in our evolutionary memory. One purpose of archetypes is to bring elements of our unconscious to the surface. Flirting may be the action of an archetype stimulating the collective unconscious. Dragonflies have shown me the deeper importance of paying attention to what attracts my attention.

For years, I focused on the mythological, the symbolic aspect of each dragonfly. As I studied what dragonflies meant to different cultures, I developed a sense of their archetypal meaning. As with most insects, they represent transformation—eggs becoming larvae becoming adults. My particular interest focused on dragonflies as messengers between worlds or souls of the dead. Now, when I encounter a dragonfly, there's the chance that something significant is occurring. I stop and ask myself if my unconscious might have knowledge for me. Or perhaps my ancestors are attempting to contact me.

As a birder, I once thought that placing too much importance on knowing the names of natural things interfered with pure observation. I watched too many intense birders care only about a name they could add to their life list. Nothing about the bird mattered once they'd properly identified it. But naming, I've learned, is the first step toward understanding an organism, which leads to exploring the role it plays in its particular part of the natural system.

I didn't need to know the names of dragonflies until one day, I did.

One summer, I began noticing different behaviors in different dragonflies. First, I focused on their wings: did a particular dragonfly perch with its wings perpendicular to its body (of the suborder Anisoptera)? Or did it hold its wings back, parallel along its abdomen when perched (suborder Zygoptera, or damselflies). Gradually I began paying attention to color and markings, size, location, and flight season. I wondered, based on my experience and belief in their archetypal and symbolic power, if increasing my knowledge and specificity might result in clearer messages from the inner world.

For a decade, I've documented and analyzed my dragonfly encounters. Most recently—last week, in fact—while walking up a steamy trail in Maine.

Rain had been falling all morning. The new sun, which found its way through thick trees, puddled in large pools along the trail. I had a lot on my mind and didn't pay much attention to the many different flying beings rising as I passed through each sun pool. The third time a group of creatures rose in front of me, I thought: wait. "Isn't May too early for dragonflies? But if not dragonflies, what are these insects?" I followed the next one I flushed into the deep grass, a risk this year, based on the current tick infestation. Definitely a dragonfly, it landed close to the ground on a dried twig. Slowly lowering myself, I got a glimpse of its tan and black abdomen, its brown eyes, and I felt hope watching its wings shimmering in the sunlight. I snapped a photo before it flew off.

Based on its small size, I guessed it was a species of meadowhawk. I stood there for a minute hoping to understand the purpose of this very significant flirt. Back at the house where we were staying, I googled "Dragonfly symbolism" and just for fun, I clicked on a random selection and came up with this: *"Getting beyond illusions . . . [the dragonfly] shows us the path to new worlds as she dances in and out of mystical portals . . . through the mists of change . . . with swift precision actions and movements.* Of course.

I opened my *Dragonflies and Damselflies of the East* book to the meadowhawk section and thumbed through it. I stopped at the photo of the White-faced Meadowhawk and read the description, which described to a tee what I'd captured in my photo: *Sympetrum obtrusum*, a female.

Without thinking about it, I had not only read the complex description, but *understood* it. "Abdomen tan with wide black ventrolateral stripe from S4-S9, covering larger parts of the segments toward rear," for example. For the first time, I wasn't intimidated by the scientifically academic language. The Metal Earth model had made the difference.

When I built my Metal Earth model I tried to follow the steps in the Assembly Flow Chart *exactly* and *in the order given.* The marathon assembly process reminded me that my experience with dragonflies had included very little anatomy. I'd ignored any point of species differentiation which required killing and dissecting the animal to find out who it was. In putting together the model, I discovered that there's much more to a dragonfly than wings, abdomen, head, eyes, thorax, and legs.

Using my field guide for reference, I went back over the instructions and labeled the parts with their anatomical names, noting interesting details. For example: Part 7, properly bent, became the box-like synthorax, to which the legs and wings are attached. This is a major part of the organism, housing the large and important flight muscles. Dragonflies, I've learned, fly with such effortless precision because each of their four wings operates individually. In the living Common Green Darner, the thorax is "bright to dull grass green," according to the book. Reproductive and digestive organs are contained in the long abdomen, its length necessary to provide the weight to balance the dragonfly. According to the book the darner's abdomen is divided into eight segments—S-3 through S-10— which are often important in distinguishing different species. An actual darner has bright blue segments (through S-6), dull, bluish green (7-8), and mostly dark (9-10). The abdomen parts are 12 through 19 in the instructions for the model.

The head, including the dragonfly's massive eyes, is comprised of parts 1 through 6. Each eye took forever to build. An eight-tabbed piece shaped like a tiny snowflake required bending into a hemisphere. I learned about the face and mouthparts by their absence: The frons and clypeus, labrum, and mandible were even too complex, too intricate for the sadistic designer who created this model.

I built my Metal Earth model one year ago. Recently, I came across the instruction sheet with the notes I'd made when the ocelli *flirted* with me. The ocelli, I recalled, was the smallest piece on the model, so small I couldn't pick it up without tweezers.

If, as Mindell suggests, "the everyday world is a huge, semi-human field of relationships, of flirts and flashes," that "everything is you," then symbols that flirt with me are my unconscious attempting to communicate with me. The flirting ocelli becomes *my* third eye. My unconscious is suggesting that I need to acknowledge my third eye chakra.

I've never paid much mind to my chakras. I know there are seven of them, that each one has a center of spiritual power. And I assumed like most things these days there are those people who believe in them and those that, since they can't be seen, don't. Learning about the ocelli becomes *proof* of my third eye, albeit a different kind of proof.

The third eye chakra, I'm discovering from the literature, is the center of insight and inner wisdom which we can trust without fear or delusion. If open and awake, our third eye opens us to being able to imagine our lives. It is the third eye which sees and understands archetypal images.

My ocelli, although invisible, are mirrored in that tiny, nerve-filled organ on the top of the dragonfly's head. It reminds me to move confidently in the world fueled by true wildness, the power of which comes from knowledge contained in my inner, unconscious, evolutionary world. This knowledge is not limited by space or by time.

The Blue Gate

STEPHEN TRIMBLE

At twenty-four, I found myself passing through what southern Utah locals call the Blue Gate. Looking back, I can see this path through spare, sere badlands as the gateway to my adulthood, the physical equivalent of my passage to the rest of my life.

I was headed into a new place, Capitol Reef National Park, to begin my season as a park ranger/naturalist. Two "gates" frame this Capitol Reef country. On the east, austere outcrops of eroded clay—the Blue Gate, where the highway rounds the fluted skirts of blue-gray Mancos Shale that drape the slopes of North Caineville Mesa. On the west, it's the Red Gate, the first flare of sandstone cliffs where the outrageous colors of the canyon country announce themselves, roiling the calm seas of gray-green sagebrush, drawing travelers deeper into the redrock wilderness, downstream toward the Colorado River.

Beyond the Blue Gate, I kept heading west, swinging around the curves and against the grain of the Waterpocket Fold, the tipped-back wrinkle of rock layers that runs for a hundred miles across southern Utah. Each mile took me into new strata, each with its own textural personality and mineral-stained flamboyance. When I reached what I soon would learn was the bentonite clay of the Morrison Formation, I remember calling out in wonderment, "My god, those rocks are *purple!*" Those Morrison hills were pretty close to magenta, which was always my favorite color in the crayon box.

My highway followed the Fremont River, a ribbon of life-affirming green in a daunting expanse of stone. Cottonwoods defined this verdant corridor, as both river and road threaded the canyon between golden domes of Navajo Sandstone. And then the Fremont canyon widened. The fractured red wall of the Wingate Sandstone lifted and pulled away from the river, creating a tiny, isolated pocket of irrigable fields with rows of fruit trees between river and cliff, a pioneer one-room school and an old Mormon farmhouse sited like a movie set within the green.

This was the village of Fruita, a cherished oasis in the labyrinth of canyons of the Colorado Plateau. This would be my home for the next seven months. In the more than forty years since I arrived at that seasonal ranger job at Capitol Reef, I've lived all over the Four Corners states, but this many-layered little haven where water meets cliff still lies at the heart of my spiritual home.

I came of age at Capitol Reef. I wasn't yet an adult, but I made strides. I was actively building what University of Virginia psychologist Meg Jay calls "identity capital," investing in experiences that would find their way into my adult self.

The slickrock landscape expanding outward from Fruita was the matrix that made possible my leaps toward maturity, a blank canvas for personal R&D, research and development into my ripening identity.

I started as a novice. Scared of scorpions, I slept on top of picnic tables at backcountry campgrounds. I drove my 1962

Dodge Dart into sandy washes and up rock-studded hills, places where such a suburban vehicle should never have been, gradually increasing in skill and decreasing the hours I spent freeing tires from the sand. I walked the park trails, photographing everything from Sego Lilies to Desert Spiny Lizards to swirling crossbedded stone. I grew bolder, backpacking into the most remote corner of Capitol Reef—Halls Creek Narrows. I grew comfortable, no longer afraid of scorpions, enchanted by Canyon Treefrogs when I shared a dip with them in potholes. I reveled in making a swath of south-central Utah my backyard.

My home base in Fruita centered these explorations. Everything becomes vivid where precious water runs through red cliffs. You plant your home next to water and range outward into the wilds. You take calculated risks, but return to green, acutely awake and alive.

Solitude in the slickrock feels different from solitude along the creek, by the lake, next to the trickle of a spring. The streams, the rivers, run with an emotional resonance that crackles with life. Near water, I feel connected to humanity, closer to my heritage as just another creature, tethered to the greater community. Beyond, down the trail into John Wesley Powell's naked "wilderness of rocks," I strive to meet the challenges of the desert wild and then return from the maze of stone canyons with new ideas, new strength, to water. Here, in repose, I can make sense out of new experience.

To grow, to mature, to become capable of relationships balances independence and empathy. Somehow the stunning contrast of green oases within the redrockscape echoes the contrasting emotional strengths of these two poles of becoming human.

I'm not the first writer to notice the power of that little pioneer village of Fruita, nor the first to return here again and again on pilgrimages.

Wallace Stegner fell in love with the place nearly a century ago. The great western writer grew up in Salt Lake City, visiting Capitol Reef country in the 1920s and 1930s, making his first trip when he was just fifteen. Stegner's nostalgia for Fruita gave him a pivotal setting for his 1979 novel, *Recapitulation*.

In this novel, Bruce Mason, Stegner's alter-ego character, takes his first love, Nola Gordon, to southern Utah for a family gathering. When the couple drives off on their own, they make a beeline for Fruita, the place where Stegner himself came of age. With his "dreaming eye," a much older Mason/Stegner remembers what happened at Capitol Reef.

"They are in a pocket of green among red cliffs. A dusty track turns off left. 'Here,' the girl says, and the driver swings the wheel.

"On that bedroll in the moon-flecked shadows of the cottonwoods beside the guggle of an irrigation ditch under the Capitol Reef, [in] the stillness that sifts down on them like feathers, [where] a canyon wren drops its notes, musical as water," Bruce Mason lost his virginity.

In *Recapitulation*, Bruce and Nola do not live happily ever after. But Mason remembers that night at Capitol Reef as "a great grateful tenderness." The place and its moment in time, so obscure and distant and remote, remained vibrant to Stegner throughout his life, as my time in that same place does for me.

My job as a ranger consisted of paying attention—and then passing on my stories and newly-acquired knowledge to park visitors. I was fresh to the wilderness, fresh to teaching, fresh to a professional role. I drew my strength and sustenance from the land, from hiking the backcountry, from reading Stegner and Clarence Dutton and Ed Abbey, Ann Zwinger and Willa Cather—and doing my best to sweep up the park visitors, timid and new to the country, and galvanize them with my passions. I was in love with the place. I was in love with *learning* about the place. And I wanted every traveler to fall in love as I had, to leave Capitol Reef bubbling with exhilaration about the stunning union here of water and rock, time and twisted juniper.

Wallace Stegner gave me one model for my work as an educator and writer. In his nonfiction, he used his experience in canyon country to take his readers into the heart of the Colorado Plateau. He came back to "the Capitol Reef" whenever he could conjure a reason, "whenever we were within three hundred miles of Fruita," the village a touchstone, both in life and on the page. He never wrote about landscape without fieldwork, without immersing himself or re-immersing himself in the place, even if briefly.

In 1977, Stegner visited southern Utah to gather material for his essays in *American Places*. Working with these fresh impressions, he captured the feel of that "lost village" of Fruita, a "sanctuary" for "enthusiasts with the atavistic compulsion to hole up in Paradise."

His research consisted of soaking up the country and then tracking down old-timers and listening attentively for color, for telling quotes. "The land is not complete without its human history and associations," wrote Stegner. "Scenery by itself is pretty sterile."

This, from the man who wrote the hallowed *Wilderness Letter* in 1960, when he mused over the view from nearby Boulder Mountain and propelled himself to that last soaring paragraph that became fundamental scripture for the conservation movement:

> We simply need that wild country available to us, even if we never do more than drive to its edge and look in. For it can be a means of reassuring ourselves of our sanity as creatures, a part of the geography of hope.

And so wilderness isn't just about solitude. The Big Empties teach us the most when we know about the people who have been there before us, from Ancestral Puebloan to Paiute to John Wesley Powell to Stegner himself. When we know about the Mormon farmers and cranky outsiders who made their homes in places like Fruita. All this, along with knowing the difference between mesa and butte, bobcat tracks and coyote tracks, blackbrush and sagebrush, gnatcatcher and nuthatch.

I discovered literature and local color and natural history and wilderness and love all at the same time, and that mix has been heady for me ever since. I made these discoveries as a college student in Colorado, as the Sixties turned to the Seventies, marked by the first Earth Day and the Vietnam draft lottery and the break-up of the Beatles and the unfortunate presidency of Richard Nixon. I managed to avoid the draft (with bad eyes), and I was appropriately fierce in my protests of the war. But my real passions ran to battling the scourge of development engulfing wild country.

I followed conservation politics. When David Brower was forced to leave his position as director of the Sierra Club—after becoming too political in his fight to save the Grand Canyon from dams—I went with him, outraged by the conservative old Sierra Club board, dropping my membership and signing up for Brower's new Friends of the Earth. I was just eighteen.

My hiking buddies and I looked for adventure, we looked for special places. We ventured into the Colorado Rockies to climb and backpack. Even more tantalizing, the redrock canyonlands of southern Utah lay just a few hours farther west. And we came to those canyons just as the great tragedy of losing Glen Canyon under the waters of the "Lake Powell" reservoir sparked a wave of elegiac regional literature.

Edward Abbey published *Desert Solitaire* in 1968, complete with his story of bobbing through Glen Canyon on dimestore rubber boats. Working with Eliot Porter's photographs of the lost side canyons, David Brower created the Sierra Club's 1963 *The Place No One Knew: Glen Canyon on the Colorado* as an epitaph and apologia for his role in allowing the Glen Canyon Dam to be built.

We couldn't go to the Cathedral in the Desert, the heart of the heart, now inundated by Lake Powell, but in my ranger job at Capitol Reef, I now had Fruita as my secure perch for exploring outward into what was left of the inner canyonlands. I began to tick off the iconic destinations I'd read about, walking to The Great Gallery of pictographs in Horseshoe Canyon, hiking through Muley Twist, rafting Cataract Canyon, exploring the magical side canyons of the Escalante wilderness.

I needed intellectual guides into this wilderness. I needed teachers of natural history, writers with a sense of place. These mentors and experiences and places shaped me, tempered me, healed me when I was heartbroken. As a slow-to-mature man, my love for place developed before my ease with intimacy, but gradually my love for mountains and mesas spilled over from land to people. If I could fall in love with Capitol Reef or the

Sangre de Cristo Range or Baja California, surely I could fall in love with a woman.

The wilds taught me, the writers helped me to articulate what I saw and felt, and I brought all of that newfound engagement to bear in my relationships.

I was on an emotional journey that took me toward adulthood through the transformative alchemy of the wilderness. The book that most intensely captured my passage was the Sierra Club book, *On the Loose.* Many of us who discovered the thrill of wild country in the 1960s remember with tenderness this small book that perfectly matched the times. Two brothers, Terry and Renny Russell, had grown up wandering in Joshua Tree, Glen Canyon, Point Reyes, the High Sierra. They had done what I had done and distilled their coming-of-age for me in this book, their hand-calligraphed testament. More than a million copies of *On the Loose* circulate out there in the world, waiting to crystallize feelings about the middle of nowhere for new readers.

I copied quotes from *On the Loose* onto the matboards that surrounded my photos from my hikes. I can recite those passages from memory, still. *Adventure is not in the guidebook and Beauty is not on the map. Seek and ye shall find.*

I didn't realize until many years later that the two Russell brothers had preceded me to Fruita and Capitol Reef. When Renny Russell published *Rock Me on the Water* forty years after his brother died in a Green River rapid, he told the backstory to *On the Loose*, filling in the family history that led to those calligraphed words that had been so influential in my youth.

Terry and Renny grew up in a California family that celebrated art, horses, music, books, photography, and wilderness. Their Aunt Elizabeth was a key influence. With her husband, Dick Sprang (a cartoonist who drew for Batman and Superman comic books), she ended up in Fruita, of all places. I knew Terry and Renny had wandered the canyon country, but, to my surprise and delight, here they were in Fruita, hanging out in my favorite village. The Russell boys spent the summer of 1958 there, "mingling with Fruita's eccentric inhabitants and merging with the slickrock wilderness surrounding the park."

They returned many times, and "left their tracks along the scant game trail leading to the summit of Mount Ellen and in the blue-gray Chinle Formation soil searching for petrified wood and dinosaur bones." They hiked down the Escalante to Cathedral in the Desert. They floated through Glen Canyon.

In 1965, when Terry tragically drowned in Steer Creek Rapid in Desolation Canyon, Renny walked forty-five miles downstream with no shoes. He came upon a picnicking couple who drove him all the way to the home of his aunt and uncle, who now lived at Fish Creek Ranch at the base of Boulder Mountain a few miles from Fruita. This had been the planned rendezvous with family after the river trip. Instead, Renny brought the news of Terry's loss.

The land between the Blue Gate and the Red Gate was one of those places that taught Terry and Renny Russell "determination, invention, improvisation, foresight, hindsight." On the loose here, they learned who they were. They created a relationship with the redrock canyons and then came back to their family homesteads, cabins shaded by groves of cottonwoods, perched at the edge of wilderness. A few years later, I, too, would pass through the Blue Gate to mold myself from the raw materials of wild country and family and community.

My identity comes right out of this heady mix of slickrock, solitude, lyrical nature writing, identification with place, and living in isolated western outposts dotted across the map wherever pioneers could find water.

In *The Defining Decade,* Meg Jay articulates what these experiences accomplish for twentysomethings. These are the years of exploration to build "identity capital," the mind-expanding encounters that adults draw on for self-definition. Jay notes that eighty percent of life's most defining moments happen by the time we're thirty-five.

For those of us long past thirty-five, this sounds a bit discouraging. But her figures remind us how critical these explorations in our twenties can be.

We can bond with a place at any age. We can fall in love with a new landscape, learn its history, hang out with the old-timers, attend to the nuances of natural history that define the personality of place. We can add whole new sweeps of the earth to our home landscape.

But Meg Jay is right. There are intervals of special receptivity that mesh with our predilections, and we are most responsive when we are young. I count the Southern Rockies, the Great Plains, and the Great Basin among the landscapes I love most, and I came to know each region when I was at my most vulnerable. But since I first came to the redrock canyons and swam through a pothole into the cradle of a slickrock slot, wooed by the cascading song of a Canyon Wren, I've never been the same.

That most powerful bond could have happened under the spell of a meadowlark's whistle and warble in the Dakota badlands. Or below Longs Peak in a sunlit glade between spruce and fir, the flute of a Hermit Thrush and the *craak* of a Clark's Nutcracker calling to me in surround sound.

But that alternative pairing of person and place didn't

happen in those landscapes with the same permanence as my bond to canyon country.

I built my own identity capital in my twenties at Capitol Reef. I pinned my identity, my life, and my happiness on a spire of slickrock. It was a good choice. I'm still here.

New Words, Lost Words, and Terms of Endearment

LAURA SEWALL

The *Economist's* 2015 end-of-year obituary was for those words deemed too difficult for testing the scholastic achievement of America's high school students. They are the disappearing and dying words. No doubt the test-makers have a point—particularly with reference to underprivileged students in underserved school districts—but surely many of the eulogized words are well known, and should be. Think of *solicitous*, *bombastic*, and *egregious*—all rather useful terms in the 2016 presidential election season, for example. In the interest of useful terms, I will particularly miss *obfuscate*, *ubiquitous*, and *amenable*. And with still plummeting declines in biodiversity, we might also appreciate having words like *incontrovertible*, *subjugated*, and *inane* on the tip of our tongue. These are honest words.

The disappearance of specific names for plants and animals is at least equally concerning. A creative series of studies show that words identifying species have been subsumed by superordinate, generic terms. In this sad slippage of language, oaks and ashes, for example, are reduced to "trees." Similarly, eiders and mergansers have become "ducks" and Ruddy Turnstones and Willets are lumped together as "shorebirds." Psychologists Altran and Medin claim that the disappearance of species-level references demonstrates a form of human devolution. They support their hypothesis by counting instances of tree names found in the Oxford English Dictionary in hundred-year intervals, beginning in 1525. The peak of references depicting tree species occurred in the nineteenth century, just prior to the onset of the Industrial Revolution. By 1925, the occurrence of specific tree names had declined dramatically, as had references to trees in general. The disappearance of such language carries additional losses, including knowledge of what grows where and why, of who nests there and when, of plant communities and soil—and so on. In a few generations, the fallout represents a radical loss in commonly shared ecological knowledge. For Altran and Medin, it is evidence pointing to a seriously maladaptive form of dumbing down.

In my way of thinking, the decline in biological vocabulary represents a baseline shift in human cognition. Shifting baselines usually refer to the ratcheting down of environmental abundance and vitality, but the loss of language is directly analogous. In the case of a shifting lexicon, the loss refers to human capacity rather than natural resources. It is a decline in the vibrancy of language, of precision and imagination, and of conceptual range. For example, "a sense of place" requires knowing who lives in the neighborhood—hooting at dusk or wandering by in the dark of night—or something about the plants that burst through cold ground or stand around giving shade, or scent, or fruit. A colleague's recent experience while teaching field methods in a salt marsh illustrates a dire loss of language with respect to simply orienting one's self in the outdoors. She asked her students to sight the horizon as a reference for determining elevations along a survey line. Several students asked what she meant by the horizon line. One student pointed to just beyond her feet, to the edge between solid ground and the incoming tide, and asked, "Do you mean that?"

If we do not know what is meant by horizon, if we have never contemplated a horizon line, how could we fully appreciate the notion of "expanding our horizons?" Surely, Martin Heidegger's provocative thoughts on what occurs within the space contained by the horizon line would be lost on us. Following Heidegger's thinking, a fundamental unity is created by the outer edge of a landscape, within which "everything merges into its own resting," "abiding and lingering," and "giving [of] itself from afar." It is a space that begins at our feet and embraces all that we see. If, as Marshall McLuhan declared, "the medium is the message," then an inclusive, unbroken natural landscape presumably signifies a form of reassurance and belonging, a sense of "being at home in the world." But could we ever feel such a sense of comfort without first looking up and recognizing what is meant by "the horizon"?

From a psychological perspective, this form of dumbing down could be cast as a crippling trajectory into a de-animated and self-referenced world, lacking in either perceived or conceptual diversity or abundance. But according to ecopsychologist Robert Greenway and a host of like-minded thinkers, our linguistic baseline has long lacked an adequate language to account for the true texture of our lived relationships with all things natural. Either way—by virtue of a long decline in a language rich with natural reference, or as a result of recent losses in basic biological terms—the fundamental cost of an impoverished lexicon is a form of collective forgetting. It's altogether (and all together) *forgetting that we have forgotten* the value of deepened experience with non-human others, essentially relegating wild and natural experience to romanticism or oblivion—a form of collective cognitive loss.

Of course, we are also gaining words: think of *virality*. For Tony Sampson, author of a 2015 book with precisely that title, *virality* refers to contagious phenomena within the technosphere, or contagion theory applied to network technology—as in social media. The defining words themselves could make heads spin, but most of us get the reference in a heartbeat. Like *snapchat*. Both words have entered common consciousness well in advance of having entre into either the Oxford English Dictionary or Microsoft spellcheck.

Trending words largely fall into social categories. In 2015, the word of the year, chosen by the American Dialect Society, was *they*, a gender-neutral singular pronoun. Other nominated words included, for example, *microaggression* and *shade*—the latter being "an insult, criticism or disrespect, shown in a subtle or clever manner." Like *they*, these words fell into the "most useful" category of nominations. There are six words in that category, none of them pointing to human goodness or the natural order. Of the twenty-nine words nominated overall, twenty-eight specify technology, social relations, or individual characteristics—as in *ammosexual*, referring to a "firearm enthusiast." Not a single nominated word designates a single thing biospheric or offers a peek at the environment.

Nonetheless, a new word in the category of environmental literacy is *plastisphere*. By (emerging) definition, the word designates ecosystems that live within plastic structures, like limpets crowded into a discarded dive mask. The word is also used to describe the over ten million bits of plastic rotating in every square kilometer of the Pacific and Atlantic gyres. This is a useful term; it's an oddly active noun designating a new concept. The word points to human habits and conjures up a slow moving, open ocean maelstrom funneling fish and trash into a downward spiral. As a concept, the graphic suggests a very large-scale notion of "downstream" + "oh shit."

Millibar is not a new word, but it jumped into the English lexicon with sudden salience, like the "freak" Icelandic storm predicted in December 2015, in part, by a mind-boggling pressure drop measured in *millibars*. The plunge in atmospheric pressure—fifty-four *millibars* in eighteen hours—was three times the criteria for explosive, or *bomb*, cyclogenesis. That phrase describes a quickly intensifying, extra-tropical storm. Like *millibars*, *bomb cyclogenesis* is not new, having been introduced in the 1950s and more explicitly defined in the 1980s—but our awareness of it might be, as we adapt to a new *Eaarth*. That term, *eaarth*, was coined by Bill McKibben in 2010 to capture the degree to which fundamental earth systems, like weather, climate, and the carbon cycle have changed. The Icelandic storm, for example, was among the five strongest storms ever recorded in the region, in the hottest year ever recorded to date.

This illustrates the rapid co-generation of language, perception, and consciousness and, consequently, the importance of our linguistic choices. Language, we know, exerts a powerful force upon perception. It sets the stage—our words and phrases sparking neurons, generating transmissions, and igniting associations. The neural activity creates a perceptual readiness, or "perceptual set"—an interpretive context—in the milliseconds before we look and see.

For clarification and as a visual psychologist, I must add that we condition our senses by what we pay attention to. With respect to language, we most easily see what we have already named (requiring our directed attention), our brains so wired and ready to zero in on whatever has been previously identified, especially with repetition and self-interest thrown into the mix. Our "searchlight" of attention naturally beams in on such a compelling mix, our neural networks cued to the signals like any self-organizing system worthy of learning and adaptation. Our neurons fire away excitedly and what resolves into focus matches much of what we expect—and what we have already named, or elaborated upon with language. With repetition (inevitably using multiple modes of attention), the synaptic connections become stronger and fire off more readily, requiring only a hint of signal, like a flash of feather.

My neuroplastic reference to the power of language in shaping perception is literal: If we don't use specific words to identify and describe what we see, it is unlikely that whatever it is would be transmitted frequently enough to build and strengthen the neural associations that give us our capacity to easily recognize relevant patterns, much less pick up any kind of adaptive advantage or telling nuance. For example, the just-noticeable difference between Bay-breasted and Blackpoll warblers in fall plumage is subtle: only a flash of feet or a conspicuous wing bar reveal two distinct species, a profound depth of relations, and an abundantly diverse world. Without specific names, the warblers would be called "small birds"—like hundreds of others—rendering the opportunity to refine our acuity, and thus our knowing of the world, lost to generalized oblivion.

I was once committed to the primacy of perception, but I can no longer meaningfully distinguish between language, perception, and consciousness. They flow together in rapid self-organizing streams of influence. Words help to bring perceptual phenomena into being and awareness; they hold phenomena in mind; and they arise out of the cultural milieu and landscapes that keep us most attentive. With our attention focused beyond ourselves, somewhere out-beyond, we are primed to wire up new knowledge and name it, soon to be followed with a wrap of conception and understanding. But, too, what we have previously perceived rests on the tip of our tongue. I'm suggesting the miracle of an iterative, adaptive mind, and our good fortune to be embedded and roaming within an equally miraculous world of wilder signs and signals, of countless voices lending themselves to the human imagination.

In thinking through the primacy of language, environmental philosopher David Abram claims that "perception always remains vulnerable to the decisive influence of language." But Abram simultaneously maintains that language is rooted in our embodied, sensory engagement with land and others. Through gestures, rhythm, and all manner of "the nonverbal exchange always already going on between our own flesh and the flesh of the world" our language takes shape, following the contours of our experience within a landscape of many voices. Reflecting a mutually co-arising relationship between perception and language, Abram says: "Only if words are felt, bodily presences, like echoes or waterfalls, can we understand the power of spoken language to influence, alter and transform the perceptual world." In my mind, this is a sexy reference to the hermeneutic, interpretive cycle that weaves together intimate relations between person and place. We may have forgotten such relations. Or perhaps we have forgotten to speak of them, or to them.

On a late winter evening, while tucked around a large stone fireplace and yearning for spring—I asked a friend how she expresses her love for the natural world. I wondered if she talks to plants like I do, calling them "honey" and "sweetheart" while tending to them. She slid lower on the couch, closed her eyes and said, "It's just so miraculous . . ." her voice trailing off. Then she opened her eyes, sat up straight, and expounded on the miracle of seeds and plants, and the way plants let us live by pumping the sky with oxygen; on the oxygen cycle itself, and on the wonderment of life. Every few sentences she'd say, "It's a miracle!" By the end, she was radiant.

You may have guessed why I asked the question. As a culture with an ever-evolving and heavily anthropocentric lexicon, I'm claiming that we are too silent in our relationship with the wilder ones. In the 1970s—prior to the capture of our attention by technology— anthropologist Hugh Brody claimed that English, unlike the Inuit language, lacked terms or concepts for intimacy with land. Anthropologist Benjamin Lee Whorf, working in the 1930s, similarly recognized the relative paucity of the English language. He found the Hopi speaking in animated terms, their language loaded with verbs and their worldview filled by the vibrant lives of non-human others. In contrast, our noun-rich and verb-poor way of speaking conditions a worldview made of static and inert objects. No doubt this predisposes us to utilitarian relationships, thus minimizing mutuality and intimacy with all the "other" relations.

We are missing both ways of speaking affectionately and loving terms for plants and undomesticated animals, for the non-human lives that trot through our own, provoking an understanding of *other*, of something very different than our singular, anxious selves. And yet we know that our engagement with the natural world offers us gifts and good health in all sorts of miraculous ways, from the breath of life to strengthened muscles and self-confidence; to the wisdom gained by knowing our place in the world, that is, by knowing where we stand and what we will stand for; to magically restoring our attention and raising our spirits by walking in the woods, or anywhere else that natural things grow. We even heal faster from surgery and trauma when plants grow in our rooms or appear outside our windows. There is no question or complex equation here. The complexity lies in the utterly unquantifiable exchanges between our sensing selves and the many "voices" out-beyond—pulsed in rhythm, blazing in light, and pressing against us like a river running by. Only missing are the human utterances, lovingly directed toward the wild ones who croak, chirp, yelp, whistle, and howl.

If the undomesticated world offers us so much health and beauty, so much glinting and swishing, and so many opportunities to fall in love each and every day, how do we—or will we—express ourselves as a steamrolling, wideband culture sweeps us further into an era of human domination, into the *ultra-anthropocene*? My question is literal. On a recent afternoon, in a fourth-floor classroom tucked under the eaves and filled with golden sunlight, I stumbled through a somewhat abandoned expression of my love for the shimmering clouds we could see, for the brilliant sky and the glorious fall season. My students looked at me sympathetically. It may be that forthrightly expressing our love for the natural world is terribly passé, left to tree huggers or those of us who roll in warm sand because it feels good. The gasping, ecstatic, and poetic expressions are apparently quaint, romantic, or irrelevant. And so it seems that we not only lack words and language but also permission and perhaps, too, the impetus to speak. Given the high degree to which social acceptance drives behavior, this is a sad state of affairs.

In the interest of all beings, perhaps we are better off looking forward, into an emerging culture. Consider the phrase "sensationally transparent"—spoken with awe and respect.

That is how actor Eddie Redmayne spoke about his conversations with transgender people while preparing to make the film, *The Danish Girl.* Surely the phrase carries cachet. Suppose it propagates forward, expanding collective consciousness. Imagine that a common reference to "sensationally transparent" allowed our bodies to openly yearn for everything arising on a suddenly warm spring day. Given the unbound mixing of sensation and perception, we might gasp when catching a glint of light, or praise the trout lilies rising up altogether in a few days' time. I'd argue that this engenders a less defended, more permeable and more welcoming-of-otherness way of being. I'd say that such a state lends itself to a feeling of being more profoundly connected and more at home in the world, more reassured and (it naturally follows) more sexy. Isn't this what love does to us?

I asked a thirty-something-year-old friend, Sam, about all of this—about the loss of biological words, the ascendency of techno-terms, and the paucity of loving words—at least for the natural world. He said, "What about *emojis?*" I thought of those that had been nominated in 2015 as "most notable" by the American Dialect Society. There were five of them. Four are positive, if you count a reference to male genitalia as that, and one was essentially neutral, referencing a sassy information desk person. None referred to nature.

Sam continued with enthusiasm: "We are a visual culture now, and we're talking about communication, not words—correct?" He spun through a huge number of small icons on his phone, naming categories like people, relationships, and animals. There were many (generic) ways of signaling nature—trees, birds, fish, insects, sunsets, moons in different phases, raindrops, and planets flitted by. There were also many hearts and signs for travel and food; there were flags, tools, objects of all kinds, and symbols. Many were upbeat and positive, and many were expressions of affection. I loved Sam's quick response but could not think of *emojis* as anything feathered or layered like a three-dimensional scene, especially with shifting daylight signaling a fourth dimension. In contrast to a richly textured and truly sensational landscape, the ubiquity of flat screens, the nearsighted condition of our eyes—our perception of depth clipped short and underdeveloped—came quickly to mind. I wondered (as I often have) if, by virtue of such screen-centric forms of communication operating 24-7, we are also losing the easy conception of depth as it relates to our intimate lives. And too, I wonder if an array of hundreds of emojis can remedy the situation. No, I thought. These are icons; they signify similarity, serving the purpose of generalization. Their function is to liken and lump things. They help us to recognize general patterns across domains. For example, an American kid and a Chinese kid could communicate with emojis; smiley faces and other basic emotional expressions are universal. But emojis are not embodied expressions loaded with nuance or provocation. They do not help us to note subtle distinctions nor do they speak, in any enduring way, to the natural world. Nor, I doubt, could they stir sensuality or love.

Sam is glowing as he speaks, beautiful and apparently not lacking in the least by virtue of his urban, screen-centric lifestyle. He agreed with me, that our hyper-privatized media distorts reality. However, our departure from agreement was his comment about the media frame for environmental concerns, that the language and stories of large-scale disruptions are also forms of capitalized sensationalism, almost hoax-like. I didn't say anything, but given my basic thesis, I should have. I should have said that we are truly drowning in our own garbage patch, in the plastisphere, and that we don't say so enough. To be sensationally transparent, I would have sat down and cried for coral reefs being blanched to death as we spoke, and for the one in five plant species that are doomed to extinction, and about our great tendency to look the other way, to change the subject. To be sensationally transparent, I would have let my despair for the disappearance of a wildly self-organizing and abundant natural world spill into tears—or maybe into some spikey, sideways comment.

Like a sharp blast of wind, I suddenly remember actor Jeffrey Tambor speaking with Terry Gross about his transgender role in *Transparent*. He said, "Your resources are your feelings, your resources are your depth . . . it gets stormy, but so what?" He has also said: "I'm all about anything that gives you more feeling and more depth . . . let the waters be rough."

Without fully realizing it, "rough" is where I began these comments on language, love, and good medicine. I had been wondering how we might find health in our relationships with the natural world when we know it to be profoundly suffering for the twisted sake of our material gain. In contemplating the question, I soon found myself suffering in tandem. I know the feeling well. It's a form of psychological pain, like despair—but not quite. It is a mix of melancholy and human shame, of foreboding loss and loneliness. It is missing the hundreds of monarchs that fluttered by every summer, and the blue mussels that gave of themselves whenever I had a special guest for dinner. I do not have a word, or words, to describe this mix of sensation and sorrow, of love and pain. In me, the feeling is flavored with rage, and equally, with a desire to honor the losses—and so, whatever I say, it must be mindful.

Like the returning tide, I have come full circle and am recalling the tenets of *right speech*, the Buddhist practice of speaking mindfully in the context of deeply knowing our interdependence with other living beings, and with a cultivated compassion for all beings. *Right speech* is one of eight steps on the Eightfold Path to Enlightenment, and the Eightfold Path is the Buddha's Fourth Noble Truth. The First Noble Truth is that there is suffering. The necessity of naming the origins of suffering is the Second Noble Truth. The Third Noble Truth is the realization that we may cease our suffering; the Eightfold Path is the practice required to do so. *Right speech* is the first step on the path.

Right speech has several basic tenets: don't lie, don't exaggerate, and don't harm others with words. These are direct instructions. Less obvious is a contextual instruction: learn to listen deeply. If you can listen to the suffering of another for a full hour, says Thich Nhat Hanh, you will have eased and healed a heart. By my hopeful extension, a healed heart is an open heart—one that may be touched by the sensible world once again, by both beauty and the inevitably rough waters of love and loss. Perhaps a truly open heart embraces the wide wonder and complexity held within one's far horizon, within the resting and abiding world so generously giving of itself, so richly available to our senses, the ten thousand things glinting in sunlight and waving in the wind. This, then, is when loving words roll off tongues and tumble into existence.

It is right to speak with love and affection, free of harm or *obfuscation*; let there be transparency. There will be no lies, no hyperbole, "lest the parade of our mutual life gets lost in the dark," pleads the poet William Stafford.

> For it is important that awake people be awake,
> or a breaking line may discourage them back to sleep;
> the signals we give—yes or no, or maybe—
> should be clear: the darkness around us is deep.

It is right to speak honestly and to choose our words with care; to listen long, to acknowledge the suffering of all beings, and to assist in unwrapping armor. It is right to sing our praises for blue skies and blue waters; for fat mackerel and salmon thrusting themselves upstream; for the Grizzly Bears gorging, winds whistling through a Douglas-Fir forest, and for all the rest. Loving like this is good medicine.

Just say it.

Serendipity, Sculpture, and Story

EDIE DILLON

One snowy afternoon at the close of November, I squatted next to the edge of a fallen kiva under the shelter of a red rock alcove in the upper reaches of a remote canyon in Utah. I was confused and saddened by the collapse of an ideal and an educational community in which my husband and I had invested huge amounts of time and great hopes for our children. Longing for solace, I asked whatever spirit remained in that once teeming place to remind me of a good way to live. How can one grow fully into the potential of the gift that is our life? I closed my eyes for this prayer and felt the hush of the snowfall enclose the ageless silence of the canyon. I imagined the people who had eaten, breathed, slept, argued, and died in this alcove, the questions asked, and answers provided. Opening my eyes: slow flakes, silence, a slight chill, gray green of juniper against russet canyon and, on the dirt square within my gaze, a fragment of ancient pottery. The white ground of the sherd was crossed and crossed again with black lines of varying width. Black-on-White, a name given to an archaeological period, also describes a dramatic value contrast, one of the classic principles of design. The pattern was painted by a hand that doubtless had many other duties—carrying water, gathering wood, grinding corn, possibly soothing a baby or feeding gruel to the sick. Yet in a square inch of clay, it manifested a miracle of artistic expression: beauty made from earth and pigment. With the precision of its contrast, the strength of its line, the firmness of the stroke, that deep black of blacks, it insisted that the act of individual human expression is important. Even when there are other pressing tasks, when there are heartbreaks and children to attend, even then, especially then, a good life responds to beauty. A good life echoes back, sings out, attempts to communicate through time, some measure of one's sense of the gift that is given.

But in a cataclysmic era, we may question the relevance of art. A painting cannot serve as a tent for a family whose home was crushed by an earthquake, a play won't float a grandmother to safety from floodwaters, a poem will not return an extinct bird to its place in the ecosystem. On the surface, it appears that art will not change the prospects for a life that has been crushed by poverty, war, or racial violence. It seems that sculpture will not save the world or anyone in it.

It is likewise tempting to dismiss the importance of spending time in nature. Listening to the song of a Black-headed Grosbeak on a morning in May—indeed, *knowing* that the grosbeak would return with its gift of song on the exact day that you heard it—will not, on the face of it, lessen the intolerance that obscures our common humanity. We do not suppose the expectation of the bud's quickening in March to stop hate from metastasizing in the culture, or expect the dandelion seed head expanding like a stem-bound galaxy on the parched ground between sidewalk and gutter to stop the black child from dying in the reeking stairwell.

But, somehow, despite the fact that it will not feed, clothe, or house us, we keep making, looking at, and trying to understand art. Art's apparent lack of functional value must belie some other worth. There must be something else life sustaining about encountering art, something about art that we need. And in this time of fascination with the virtual, when adolescents in darkened bedrooms and basements hunch blue-lit over video games to practice detached brutality, we are still inexplicably bound to nature. The first snow, the windy day in April, the panda cam, the glimpse of moon through fast moving clouds, the Facebook video of a sea turtle beginning its journey—all stir something both restive and vestigial. It starts as a small feeling in the core, a story remembered, a way back to mystery.

I am a sculptor, my medium "found objects"— things I find in thrift stores, garage sales, along trails and beaches. The things are often rusty, frequently rock hard, sometimes actual rocks. I use fabric, rubber, glass, interesting branches, guitar strings, old roller skates, lace. A couple of times I've been inspired by tar picked up from tideline. I spend my out-of-studio days finding things, and in-studio days solving problems within an aesthetic. The work is literally a balancing act—if I want it to look x way, I need to figure out how to connect and hold y and z. Much of the trick is, unglamorously, knowing glues. Some of it knowing tools and screws, and weights. But, way more than some artists might divulge, the process involves being led rather than actual doing. My best work happens when I follow a little not-just-me-talking voice in my head that tells me what of the many things I have pack-rat piled on tables and shelves would work just right, exactly now. And then I listen for which glue, which screw, and how. With implausible frequency, I find on some dusty thrift store shelf, or roadside, or in a box sent by a friend, just the thing I need exactly when I need it.

Most people have experienced "finding" moments and sudden, apparently out-of-nowhere answers. My eighty-three-year-old friend, who should know from a lifetime of improbable coincidences, calls them "meant-to-be's." The nature of my work bids me pay special attention to happy accidents. Serendipity is the guiding twinkle of my creative process.

Long before I made sculpture, I sought to understand the world through the natural sciences, and I hoped to help save the world through sharing scientific information in such a way that my audience would be inspired to act on behalf of nature. My first real job and a career I still hold as righteous, not to mention fun, was as a national park ranger.

The world view of the artist and the environmental scientist are each based on active imagination, an imagination not springing from fantasy, but grounded in a close attention to specifics which allows the practitioner to both see what is there, and conceive what is not.

As a naturalist when I hear the downscaling song of the Canyon Wren, I recall peppery Seepwillow scenting a light breeze, the many voices in the rushing creek where I last heard the wren, footprints of heron in wet sand. The artist in me wants to give visual expression to the joy of being in a place where Seepwillow smell and water sound and ancient covenant between heron and fish are stitched together by the song of a small bird. Both approaches use the ground of sensory experience in a process that, to paraphrase Marcel Duchamp, makes the invisible visible.

I tried first to think about this essay as a paean to the practice of natural history as it contributes to my life as an artist. In that version of the story, time spent at natural history would be sort of like showing up at a mental Pilates gym where attentiveness—a foundational attitude for the artist—strengthens its core. W. B. Yeats said, "*The world is full of magic things, patiently waiting for our senses to grow sharper.*" The naturalist and the artist both endeavor toward senses sharpened to the magic things, we rely on the alchemy of sensory delight. We notice things and we make something of that, be it metaphor or theory.

But Pilates or parallel play are not really the nut of the matter.

The practice of art and the practice of natural history are not just parallel pursuits, and one isn't merely preparation for the other. Art and natural history are nurtured and grow together from the very same root; a sense of immanence—that is, the awareness of the spiritual world permeating the mundane, and a feeling that the divine encompasses and is manifested in each detail of the material world. My artist self and my naturalist self are nurtured in the same belief—that spirit is fully alive in the world, is always shaping toward beauty, transcendent wholeness, and stories with unexpected endings.

Natural history is the evidence. The creative voice is the evidence. What is required of us is a willingness to cultivate serendipity.

Is there anything in the natural world that does not express with exquisite and terrible and intricately fitted loveliness? Blake's verse rises to mind from childhood's memory:

> Tyger Tyger burning bright,
> In the forests of the night;
> What immortal hand or eye,
> Dare frame thy fearful symmetry?

A 2005 study of European inventors found that fully fifty percent of patents resulted from a serendipitous process. Thousands of respondents said that their new ideas evolved when they were working on something unrelated. A typical example: endocrinologist Dr. John Eng noticed that certain lizard poisons damaged the pancreas. He ultimately found a compound in Gila Monster saliva that led to a treatment for diabetes. An associate describing the seemingly random discovery noted that Dr. Eng discerned "patterns that others don't see."

In the 1960s, Gay Talese declared that New York was a "city of things unnoticed" and set out to notice them. He encountered a colony of ants at the top of the Empire State Building, followed the wanderings of feral cats, and cataloged the providers of shoeshines. His book *New York: A Serendipiter's Journey* cataloged the discoveries and gives us a term for someone whose senses are sharpened for delight.

For several decades Sanda Erdelez, a University of Missouri information scientist, has worked to find out if we can train ourselves to be more serendipitous. In the mid-1990s, her research focused on discovering if different people interact with the world distinctly in ways that foster or suppress serendipity. The one hundred subjects fell into three distinct groups. She called them *non-encounterers*, *occasional encounterers*, and *super encounterers*. As if watching baseball through a knothole in the fence, non-encounterers see a tightly focused world, not the whole game. Occasional encounterers might stumble upon and recognize moments of serendipity. The super encounterers reported seeing happy accidents everywhere. Here is a story of a typical happy accident in art making.

I find contemporary angel art trite, a little ridiculous, and faintly irritating. But, having never before or since thought of creating anything to do with angels, I made a sculpture called *Angelus*. The body is a piece of Precambrian granite from the mountains near my home. I saw the stone lying on a trail and immediately thought it looked like a torso that had been knocked down or fallen from place over the ages, like an ancient Greek sibyl. I made wings from two different vintage silk scarves, hand sewn and stretched over a metal frame. Along their lower edges, to catch the light and give a sense of the mortal, I stitched a row of dangling glass beads and fish vertebrae from a necklace made by a Seri woman. But I still needed a metal piece to connect silk wings to stone back. I was thinking that maybe its shape should be the elongated triangle of a moth or butterfly body. Finding no good solutions in my studio stash, I forayed to the basement of the thrift store in town that has the best boxes of garage junk. I crossed the ill-lit room to a shelf piled with dusty boxes filled with nails, greasy screws, old files, blunted drill bits, rusty saw blades—an unglittering Golam's pile with no ring in site, the tool bench detritus of a dozen deceased grandpas. A girl has to start somewhere, though, so I reached in.

Without even the beginnings of a satisfying, nail dirtying rummage, my hand landed on a brick pointer, a piece of metal shaped into an extended triangle. This one had the word *Angelus* incised onto it. Angel. A company named Angelus that made brick walls put their name on a moth-body-shaped brick pointer that ended up in the basement of Tattered Treasures in the very first box I touched. What are the chances?

"Creativity is another form of open space," Terry Tempest Williams writes in *When Women Were Birds*, "whose very nature is to disturb, disrupt, and bring us to tenderness." Super encounterers, serendipiters, pay attention and see patterns.

Awareness of serendipity is not dumb luck; it is a developable quality of mind that rises from openness to the unexpected. It is alertness to Blake's fearful symmetry, a celebration of sacred surprise, tenderness toward delight. The gift of serendipity is that we can use it to create a new story, or, in our current predicament, a new ending to the same story.

The accepted narrative of our times, a trajectory of greed, despair, and planetary ruin, is in desperate need of deflection. It is the moment for a global version of the *Choose Your Own Adventure* books published in the 1980s and still popular, where the adolescent protagonist gets to make choices that determine the plot's outcome. It turns out that super encounterers get that way in part because they expect serendipity. Appreciating that they will gain new perspective, they consciously develop a capacity for noticing, for imagination, for alternate stories.

What we pay attention to, and how that affects what we choose to keep or throw away, be they materials, ideas, or specifics of the natural environment, can change the stories we tell and the possibilities we allow. With sculpture, I engage our human and environmental predicament using the incongruent beauty of the sacred found. Awareness of serendipity—that fearful symmetry in the world full of magic things—can be extended by paying attention to the immanence in the ordinary details of the world. There is no better workshop for developing this capacity then a walk in the woods, a stroll in the desert, a gaze over the ocean, or a bug's-eye view achieved by lying on the ground in a city park.

Bird watching is a sure way to extend awareness of serendipity. You never exactly know when that singing packet of heart, muscle, and bone will appear, but you prepare with map and field guide and are ready for the miracle. People who watch birds know that small winged bodies can stitch together distant geographies by fantastical migration. Extending serendipity stretches the imagination. I imagine myself into the eyes of a migrating warbler and sense the contours of the land below, what it would be like to walk it, the feel of the shade of its trees, the smell of its watery places.

I live in Arizona, but have spent substantial time watching birds in the Carolina Lowcountry. When we have walked on a place, we can never think of it as just a shape on a map, but a fully detailed and nuanced landscape.

I can picture the pines and the oaks and the grasses and the fields and the low spots and the crabs and the high spots and the turkeys, and the herons and the warblers and the shrikes and the hawks. I can picture the dusty road and the trailer park and the church and the silent pews and the basement room and the angry boy and the loaded gun. What if the boy had opened the trailer door and found a singing bird he knew by name? What if he had expected its return and recognized the sacred in its journey and his own?

I spoke of finding new endings to stories. Many people have a dog story. Here is mine. During the years when our children were growing up and we lived next to the Granite Mountain Wilderness, the feathery red tail of a dog is always in the picture. It is most often drawing enthusiastic circles behind a compact body hopping along a trail a few yards ahead, or whapping the floor as one of the family came into the room, or balancing the funny duck and turn, full-body smile that we called Violet's "zigger dance."

The story of her arrival is legend with us: how I first saw her eight miles deep in Havasu Canyon, one of many starved "rez" dogs that my friend and I passed while backpacking, how she somehow got to the parking lot ahead of us—we dragged our sorry selves across the asphalt filled with dusty trucks and found her sleeping in the shade of my car. Violet's story with us begins with the salami that lured her onto the back seat, the copious dog barf, the chin that rested on my shoulder as I drove the desert highway home.

I named her in the car as the sun set. Violet, for the wild flower: sweet, resilient, optimistic, persistent. Thirteen miles scrabbling out a hot canyon on a shattered elbow most likely caused by human abuse, hurtling along in a smelly car at a speed completely foreign to her canyon life, and this dog chose trust.

Within a month of the surgery to remove her leg, Violet was climbing up a narrow canyon with us to a pool beneath a giant boulder with pictographs—a great spot for tree frogs. One of my favorite mental images is the deeply satisfied look when her body was fully submerged. Her head would sort of float the surface with a giant grin, completely expressive of the rare pleasure of a soak in dry country. More than once, she inspired me to take the plunge into the round swimming hole way up that canyon.

The only things that Violet truly feared were brooms, vacuum cleaners, and snakes. One day when the kids and I were plunked down picnicking on the coarse canyon sand, I idly picked up the shed skin of a Bull Snake. Violet woofed several times, meaning unmistakably: Put That Nasty Thing Down! Then, in spite of longstanding aversion, she ducked to my hands, grabbed the skin, and hopped at a fast clip down the wash to deposit it where it couldn't hurt the dumb, lovable people.

We moved into town when the kids got older and the frequent necessary drives back and forth to our home by the wilderness became a hassle. The last years were hard on Violet's body. Her one front leg took the brunt of each step; she hurt from arthritis, much, if not all, the time. With limited motion, her exuberance extruded in jags of unjustified barking. But she played in the snow a little and zigger-danced when our son came home for a Christmas visit.

In May we walked with our grown-up kids to leave Violet's ashes at the roots of a juniper above the pool at the top of the canyon where years before she had saved me from the snake skin. The next June, Granite Mountain burned.

One day last spring, I went to keep company with our old neighbor and to take, at her suggestion, a walk out onto the mountain. I was reluctant; memories of the good young time— the kid days, the tail wag walking, picnicking in the sand, tree frog boulder pool days. Besides, the fire on Granite Mountain was the last one the nineteen young men on our city Hotshot crew fought and quelled. They walked away from the wash just hours before an even greater conflagration at Yarnell Hill took their young and laughing lives. The Prescott Hotshots saved the residential area next to the wilderness, but I had heard that all of our side of the mountain was ash and rock. I was flat out afraid to walk toward all that sorrow, did not want to invite the loss.

The creek was running high—we hopped across on rocks rolled into place for the kids years before. We walked along the old trail, between the Cliff Roses, bee-filled and buzzing, as always, and out to the one giant Ponderosa whose shade was always the first snack stop. We had named the big pine Grandpa Tree. The Grandpa Tree was burned. Not a little. Completely. In place of the red column of trunk and broad feathery crown, stood a huge black skeleton, rooted to the bare ground and towering grandly, massively, magnificently sculptural against the cerulean Arizona sky. Forty feet above, Acorn Woodpeckers flapped and squawked in a lively aerial dance. A neighbor who stubbornly didn't evacuate during the fire said that the massive tree had "gone up like a Roman candle." Right on, Grandpa, I thought, that is one hell of a way to leave your life—with a blast and a boom to become a monument and home to families of red-capped acrobats.

We walked along the mountain's base where the hillside piñons and oaks had sheltered jackrabbits and quail and obscured the shape of the land with their bunched and rounded green. Now the forest looks like first stages in an instruction book for drawing trees—all stark lines and angles. As their blackened bark falls, the trunks and branches turn ghostly silver. We see that after the fire, the true contour of the landscape is revealed—a series of small mounds rising toward the crest and covered with waving gold grass.

We climbed above the little narrow wash and further up along the side of the canyon and the place where the round swimming hole used to be. We clambered up the big angled rock, the place with the tree frogs and that boulder with the spiral pictograph. We finally got to the flat spot above where grew the Alligator Juniper where we buried Violet's ashes.

The juniper burned way down beneath the roots. There was so much runoff after the fire that the water's force cleaned out everything to about a foot below the former ground surface. The roots were left silver and so smooth. Violet's ashes had washed away, disappeared downstream, gone with that determinedly joyful life. Nothing physical to show that she ever was, or ever was a part of us.

We stood a while in silence and looked out across the valley to the San Francisco Peaks far to the north and covered with late snow. Before the fire the trees had obscured the far horizon, now a heart-expanding view of fifty miles.

We crossed a small rise and descended a crease in the hills, a parallel canyon. And the little canyon opened to a small wet meadow. There we stood in a sweet green bowl, last fall's yellow stems and seed heads bobbing lightly in the breeze. The ground was unexpectedly spongy. We looked to our boot toes and saw the ground at our feet thickly covered in tiny purple flowers, each bloom about the size of the nail on my little finger. There were hundreds, maybe thousands of them. Violets. Sweet, resilient, stubborn, optimistic.

In knee-high yellow grass, among the silver trees, with her flowers at my feet, I could almost see Violet hippy-hopping along with us, taking delight in a pool of water, and surveying her good world with nose to the breeze.

This day was graced with serendipity, an ordinary thing gone magic, a new and different ending to a sorry tale. We know that young men die and mountains burn. We know that life is full of sorrow and change, and we sometimes find hope that change brings new beauty.

We know that the ashes of a dog, however beloved, washed down canyon by fire and flood do not become the seeds of flowers.

But this is a true story.

Spark and Fire

SARAH JUNIPER RABKIN

I remember perching one morning on the lap of my kindergarten teacher, Mrs. Cocklin, boosted just high enough to reach the surface of her immense oak desk. There, atop a lesson-plan blotter pad, sat the gift I had made for my parents: a folded note on stiff white cardstock, its cover adorned with dried leaves pressed between trimmed sheets of waxed paper.

The card lay open beneath the tip of a green marker that I clutched in my right hand. Under Mrs. Cocklin's watchful eye, I was supposed to write a greeting on the blank page inside: "To Mommy and Daddy from Sarah."

I had learned how to form the words, but I wasn't sure where to make my first mark. There was probably a correct location, and I should probably know what it was: High in the upper left corner of the page? Somewhere in the middle? I began to draw my first shaky letter—then reconsidered, moving on to a possibly more acceptable spot.

The naptime classroom was deathly silent. The rest of the kids lay supine on vinyl-covered mats on the linoleum floor, awaiting their turns to sit on the teacher's lap and sign their own cards. They held their arms rigidly at their sides, as Mrs.

Cocklin always insisted we do during rest period, lest our hands entwine with those of our neighbors or, worse, wander to forbidden realms of our own bodies.

My hand hovered over the desk, hopping several more times from one position to another on the cardstock, until the page was stippled with little green jots marking my false starts. I considered asking for help, but admitting confusion seemed more shameful than just mutely fumbling. Finally, Mrs. Cocklin reached around me from behind, stilled my searching hand in her bony one, and chastised me in a rasping whisper: "What are you *doing*?!"

What I was doing was trying very hard to get it right— and, in the process, getting it all wrong.

I learned in kindergarten, and in much of my formal schooling all the way through college, to care above all about getting things right. This often meant hiding the process I went through to arrive at new knowledge—concealing that journey not only from my teachers but also from myself. In short, I never really learned how to learn.

If my grandchildren's experience is any indication, kindergarten classrooms have become livelier, more encouraging places since I sat on Mrs. Cocklin's lap. Yet even now—especially in the current rote-learning and test-crazed atmosphere—too much institutional schooling in this country seems destined to undermine kids' faith in their own curiosity and resilience in the face of failure: the very qualities that make human beings good learners.

What I've happily discovered in later life, though, is that no matter one's age, there are opportunities to limber up the mind and reclaim a fruitfully freewheeling approach to learning. I know no better tinder for rekindling joy in the process than "nature study"—an old-fashioned term, and one I like, that stands for the array of activities and attitudes often referred to now as "natural history practice." By whatever name, the heart of this pursuit is curiosity, humility, and loving attentiveness. It is recognizing the vast and vital world of interacting beings and cycles, of the biosphere that gave rise and gives sustenance to us humans—we who are ourselves no less the planet's creatures, for all our earth-changing powers. Nature study is the desire to understand something of what's going on in the living world, a style of inquiry that can lead down an endlessly interesting path. Pursued in solitude, it's enlivening; in the company of others, it can foster kinship and generosity.

🌿 🌿 🌿

When prominent thinkers at the 2015 Aspen Ideas Festival were asked how people become better learners, their responses were remarkably consistent. Anne Libera, a teacher of comedy studies at The Second City, believes the biggest impediment to learning for students of any age is the "need to be *right* first thing." Similarly, journalist Amanda Ripley, author of *The Smartest Kids in the World: And How They Got That Way,* holds that Americans will learn more effectively when we develop greater acceptance of "the discomfort that comes from not knowing."

Industrial designer and corporate CEO Tim Brown thinks we need to cultivate the ability to pose our own questions—especially ones that challenge assumptions or reframe problems. And Stanford University math education professor Jo Boaler cites studies showing that once students get comfortable with errors, they not only enjoy math more and do better at it; their brains operate differently. "When they make a mistake, their brains actually spark and fire," she says, "because they know mistakes are good."

Mistakes are good. I've heard versions of that maxim countless times; intellectually, I know it's true. Mistakes can be fertile, pat answers sterile. For me, though, putting this insight into daily practice requires unseating deep habits of mind. Nothing opens the tight fist of my thinking like unstructured hours sitting by a mountain stream or walking a wooded slope, taking curious notice of local goings-on—or sitting at the kitchen table day after day, watching the hummingbirds that zip and dart around the feeder outside our window. Why have the emerald-and-magenta Anna's hummers, until last week the lords of the realm, suddenly vanished, replaced by little rust-and-olive Allen's? What causes the interlopers to tolerate the returning Anna's a few weeks later, feeding alongside them—until one day the males of both species launch into furious aerial skirmishes?

Over the years, some fine naturalists have helped me build such questions into launching pads for learning. One is California natural history evangelist John Muir "Jack" Laws, whose autodidactic explorations of neuropsychology inform his impassioned teachings on "thinking like a naturalist." If you take a class from Jack, he'll send you into the field, journal in hand, to "shake up your mental Etch A Sketch." In lieu of the common exhortation to "look hard" at an object of attention, he'll have you begin with an unfiltered litany of observations: *everything you notice* about the creature or phenomenon you're observing. Next, you start asking questions: what do you *wonder* about what you're noticing? And then you move on to every *association* that arises, no matter how wacky or irrelevant it might seem: what does your object *remind* you of? If you find yourself inventing tentative explanations for the phenomena you're observing, you can begin to ask, "Could it be . . . ?"

I balked, initially, at Jack's next piece of advice: to speak those observations, questions, and associations *out loud*. Self-conscious about talking into the air, I figured maybe I could just skip the vocalizing. But when I pushed past my resistance, three amazing things happened. First, within less than five minutes, I was picking up more information about my object of study—an American Robin—than I'd ever noticed before. Second, talking out loud helped me retain a surprising amount of what I'd witnessed, even after Jack took our group through a brain-teasing puzzle intended to divert our attention from our field observations. And, third, the timbre of my mind and body was noticeably altered, as if I'd just completed a series of sprints or embraced a lover. I was tingling with curiosity and enthusiasm. My usual preoccupation with performing well had vanished, replaced by an energizing combination of humility and confidence. I wanted to keep watching that bird. I wanted to stay outside for hours.

Relinquishing the need to see myself as smart and knowledgeable feels like letting go of a debilitating illness, something toxic in the blood and heavy in the bones. Jack Laws insists that it's okay to stumble and fumble in the pursuit of understanding. Such encouragement to experiment and get things wrong is in fact more than okay: it's a balm, a tonic, a potent remedy. The heady feeling our sessions afield engender, Jack tells his students, is "your brain on nature." "It's an honest way of making meaning," he says. "It's delicious fun. If you do this with your naturalist homies, the world will become a richer, more interesting place. You'll never want to come inside!"

I've often found that opening up my senses to the intricate mysteries of birdsong or bug behavior feels very much like falling in love. Jack acknowledges this. "What is love, after all?" he asks. "Sustained, compassionate attention. It's this love that fires our stewardship, and it's more powerful than avarice can ever be."

❦   ❦   ❦

Giving an entity our sustained attention can teach us not only to know it more deeply, but to love it more fully. And the converse is also true: we learn with special alacrity when we engage in activities that have captured our hearts. Freer in my older age to choose the objects of my study, I revel in the discoveries that bloom whenever I make a point of following delight.

When a field sketching class twenty years ago introduced me to colored pencils, I was instantly drawn to the buttery feel of pigmented wax marrying toothy paper, the subtle blending of layered colors. The first time I applied my nascent skills outside of class, I used my Prismacolors to sketch an empty crab claw I'd found on the beach near my home. The claw's shaft was dotted with saffron freckles, the serrated pincers edged in shiny black, the underlying surface shading exquisitely from lemon to ecru to eggshell. As often happens when I allow myself to be tutored by the subject of a drawing, I gradually let go of finicky preoccupation with the limits of my ability. I melted into the experience of channeling shape, texture, and hue; I lost all sense of time passing, save for the intervals marked by crashing ocean waves.

I still love that journal page, partly for the three graphite-pencil contour sketches that surround the fully rendered claw. Awkwardly drawn, out of proportion—false starts, perhaps—my series of tentative outlines warmed me up for the final image. I opted not to erase those initial attempts, choosing instead to honor the record of my learning process. The rough sketches remind me of something a naturalist friend once said, affirming the great relief she feels "when I realize that I get to fail, again and again." A framed color photocopy of the crab-claw image now hangs in our bathroom, where it recently caught the attention of my five-year-old granddaughter. "I like that picture," she proclaimed. Pointing to the gray sketches, she added, "Those parts show you how to draw it!"

On a trip to Finland for an expatriate nephew's wedding, I found myself enchanted by the cadences of his newly adopted, very foreign tongue; after returning home, I decided to try acquiring the rudiments of Finnish. Conversing freely with instructional CDs alone in my car, I discovered a new penchant for language learning. When native speakers encouraged me to converse with them, though, I became painfully self-conscious. I hated coming up against my own limitations and revealing them in public. Attached to feeling accomplished, I repeatedly avoided opportunities to try actually *communicating*, which was in theory the ultimate goal of all my practicing. But an expert language learner is an expert fool: the more willing she is to relinquish pride and ego, the more fluidly and rapidly she learns. My Finnish project is in part a meta-lesson in gaining comfort looking stupid—so that I *can* learn.

I took up taiko drumming two years ago because of the thrill that shot through me every time I witnessed a live performance—a feeling akin to the one that spoken Finnish arouses. A unique combination of music, dance, and martial art, taiko is contained wildness, disciplined ecstasy, studied abandon. Inspired by ancient East Asian village drumming traditions, reconceived as ensemble performance art in the mid-twentieth century, it has rapidly gained popularity in Japan and abroad. I, too, wanted to become a vessel for the sheer life-force—*ki*—that emanated from the drummers' rhythmic sounds and graceful movements and that shone from their eyes at concerts and festivals. Since I began studying taiko, the excitement of channeling that energy has propelled me through the periods of frustration that are an inevitable part of learning a subtle and exacting discipline.

A few months into my taiko initiation, returning home after a week's teaching in the mountains, I picked up my *bachi*—thick wooden drumsticks—and revisited the borrowed practice drum that stood the corner of the living room. Before leaving town, I had been stuck in a hard patch in my taiko learning, all uncertainty and self-consciousness. Now, as I began to play, a potent wave of *ki* animated my spirit. I felt redeemed, alive, strong, centered—a clear and unobstructed channel. I realized: *this* is how you power past ragged rhythm to surefire precision: you become a vehicle for the energy, you feel the potency in your body and the need to express it through the drum. I haven't yet been able to enter this state as a matter of will. Some days the tiger is waiting at the cage door, muscular, hungry to pounce, every cell and fiber ready to act; sometimes it holds back. I keep playing, and maybe someday the tiger will spring whenever I call.

🌿　🌿　🌿

Field sketching, Finnish, taiko: not the disparate endeavors they may seem, but linked by their power to engage my full being—body, mind, and spirit. This they share with nature study, and its constant reminders of the fibers that connect my small self to a larger world. Whatever the subject, learning is never solely a cerebral endeavor, but also an embodied one—in part because our brains are not just in our heads. Some ancient cultures have long intuited this, locating the seat of understanding outside the skull. As author Philip Shepherd observed in a 2013 interview in *The Sun*, there's a Chinese word for "belly" that means "mind palace"; in Japanese culture, the seat of understanding is located in the *hara*, the energy center in the abdomen. Where English-speakers might declare that someone "has a good head on her shoulders," the Japanese say she "has a well-developed belly." Shepherd argues that this difference runs much deeper than semantics. It points to our culture's failure to recognize a major component of our own intelligence, "so we are stuck in the cranium, unable to open the door to the body and join its thinking."

Western neuroscience has been slow to assimilate this understanding, despite the observations of its own twentieth-century practitioners. In the early 1900s, Shepherd notes, American anatomist Byron Robinson described an "abdominal and pelvic brain"; in the 1920s, British physiologist Johannis Langley described not only the sympathetic and parasympathetic divisions of the autonomic nervous system, but also an independent network of visceral neurons that he referred to as the "enteric," literally intestinal, nervous system. Some of his language may be familiar from high-school biology texts, but the gut brain never made it into the books; it was too far out for Langley's colleagues, who could see only what they already believed. Contemporary neuroscientists, however, have finally begun to investigate the network of nerve cells lining the human gut, to designate it an "enteric brain," and to recognize that its role extends beyond digestive functions to our capacity for awareness and discernment. Western thinking now acknowledges a neurological basis for the intuitive perceptions we refer to colloquially as "gut feelings."

From an early age, I was taught to "rest" with my arms rigid at my sides and to approach new ideas using a fraction of my learning potential. I got into the habit of repressing my gut's thinking—palpable hunches, thrills and discomforts that

arose in response to texts and lectures, the assumptions underlying their contents and the manner of their transmission. I
assumed that my private, inchoate ruminations were beside the
point. I understand now that those musings arose from the
fertile depths of my intelligence. Had I found the confidence
to pursue those inklings, they might have inspired and fortified
my studies.

Nature study gives me room as a learner to stretch out
into gut brain and cerebral mind, body and heart. Sketching
a Lodgepole Pine sapling, I feel its branches' curving gestures
as if they were my own. Drawing the insects and other invertebrates that shower from gently shaken limbs onto a sheet
laid below, I meet the tree's micro-habitat. Filling the mar-
gins of my notebook page with observations and questions, I
begin a process of inquiry both intuitive and systematic. My
body-mind seems to learn especially well when I'm outdoors
doing this sort of thing among living and growing things, clean
air and wild water. Recent research supports this hunch, with
studies demonstrating that wilderness immersion improves
performance on tasks that require creative thinking. Even a
stroll in an urban park can improve short-term memory and
problem-solving skills.

Outdoors or in, contemplating the natural world provides
endless practice in the art of posing questions. On my study
wall at home I have photographs cut from two different magazines. One clipping depicts a gray-and-white Koala embracing
a slanting upright eucalyptus branch, in the classic sleeping-
Koala pose that humans find adorable. The side of the furry
animal's rubbery black oversized lozenge of a nose is pressed
against the trunk; its black-clawed fingers and toes are splayed
on the bark; its pink-rimmed eyes are shut into short black

hyphens, the face as a whole bearing an apparent expression of supremely satisfied repose.

Facing the marsupial in a visual echo is the image of another creature, clasping what appears to be a similar branch that slants toward the Koala's eucalyptus limb. The second animal's body matches the color of the mammal's fur, but bears only a few strands of hair; its otherwise naked, armor-like surface is textured like a thumbprint. And the legs that grip this branch are more numerous than a Koala's. It's the much-enlarged image of a microscopic koala fur mite, *Astigmata koalachirus*, wrapping itself cozily around a single strand of its host's fur. The caption says the creature's front legs are "specially adapted in size and reach to cling to Koala hair."

I glance at that pair of images several times a day, letting them remind me how weird and wild the ecosphere is, filled with ingenious adaptations and mysterious symmetries. How is it that this tiny invertebrate looks so much like a microcosm of its mammalian host? Why does its dorsal covering look uncannily like a thumbprint? I wonder what new revelations I might encounter if I were to pursue such questions.

❦ ❦ ❦

Throughout adulthood I've taken and taught dozens of field-based natural history classes and workshops, learning alongside other enthusiasts of all ages. Again and again, I've found that people who seek to be tutored by the land tend to be unusually congenial and cooperative. There's something delightfully humbling about the complexity of the un-built world, where each revelation about the intricate survival strategies of a particular organism or dynamic interdependencies within an ecosystem reveals further mysteries. Intimate encounters with the world that made us can have a damping effect on human egotism and arrogance.

The best natural history instructors exemplify this propensity, modeling not only deep knowledge but also profound humility and generosity. I think of my friend and ornithology teacher Breck Tyler—the way he attends with deep respect to every one of his students, much as he listens to the birds we observe in the field. He always made me feel as if he was learning something from the questions I asked.

The more time I spend hanging out with groups of "nature nerds," the less compelled I feel to demonstrate in daily interactions how much I know or how right I am. I have become more comfortable listening to others, trying their perceptions on for size. Given that communities of naturalists learning together often exemplify such sorely needed civility, I wonder whether group natural history inquiry on a widespread scale might nudge us significantly toward an essential social healing. There's a lot of talk these days about political polarization and a desperate need to engage in dialogue across differences. Perhaps turning together to the larger-than-human world, in a spirit of inquiry and respect, can help cure us of the impulse toward righteous disdain, help us make room for other beings in our minds and hearts.

🌿 🌿 🌿

A team of French scientists has been studying the world's drought-ravaged trees by listening to water-starved wood in a lab. They have learned that while you can't necessarily tell by looking that a tree is dangerously dry, if you put a microphone to it, the tree can tell you. Periodically you'll pick up little hollow knock-knock sounds from inside the trunk. Or that's what they sound like if you slow down the frequency a thousand-fold to bring it within the range of human hearing.

What you're hearing are tiny bubbles—breaks in the column of liquid traveling up the tree's circulatory system—that form because the roots just can't absorb enough from the desiccated ground to draw up a continuous stream. It's as if the canopy is sucking at a straw, trying to get at the film of water left on the bottom of the glass. Cut off from the wellsprings of our own confidence and curiosity, people, too, are prone to parching. May we cultivate ways of learning that draw up nourishment and unfurl new leaves to the sun. May our body-minds spark and fire; may our curiosity blossom and bear fruit.

Nature. Love. Medicine. Healing. Reciprocity. Generosity.

MITCHELL THOMASHOW

There are four encompassing challenges that summarize the human condition as we career through the early days of the Anthropocene. Challenge One is the rapacious exploitation of the biosphere and its life systems, what I'll describe as the human/nature conundrum. Do we choose sustainability or ecological deterioration? Challenge Two is the increasing disparity between rich and poor, or the equity conundrum. Do we choose equal economic opportunity or materialism and selfishness? Challenge Three is the simultaneous integration and separation of global cultures, or the tribalism conundrum. Do we live as a harmonious integrated humanity or as isolated, fearful tribes? Challenge Four is the egregious use of violence, weapons, and oppression, or the social behavior conundrum. Do we settle our differences through mindful deliberation, compromise, and service, or via conflict, extreme behavior, narcissism, and entitlement?

These challenges aren't unique to the Anthropocene, and there are many ways to frame them. Ever since I was a young boy in the 1950s I've been exposed to or experienced these issues. Throughout my entire career as an environmental studies professor and then a college president and sustainability consultant, I've worked with students, staff, and faculty for whom these issues are the source of their overriding concern. They are so big and so "classic" to human history, that they seem to take a life of their own, to be inaccessible to political agency, to require incomprehensible systemic and behavioral change. Often we seek to address them by living virtuously, as individuals and in community, reflective about our actions, self-critical to our hypocrisies, casting judgment on our inadequacies.

However classic these challenges may be, the accelerated pace of global environmental change catalyzes the dynamic interactions of our problems, demanding urgency in the face of ubiquitous threats. We must respond even though we may not be entirely sure how to do so, or whether our actions and analysis are effective. Does it really matter that you take great pleasure in observing the natural world? Does living a sustainable lifestyle significantly change anything? We can rationalize our choices by attributing virtue to our actions. But we are still daunted by the enormity of forces beyond our control. This is, at the very least, an existential challenge, provoking us to think about meaning, purpose, and agency.

For most of my career, I've assumed that deeper awareness of natural history cultivates an enhanced appreciation, respect, and even love for the biosphere. I've written books, developed curriculum, and built academic programs that aspire to apply that assumption. The "Nature-Love-Medicine" word chain is a compelling prescription. It's been a foundation for how an entire generation of educators thinks about environmental stewardship. In this essay, I'd like to expand that word chain so that it's more complete and fully encompasses the four challenges of our times.

The first word that comes to mind is Healing. I'm intrigued at how two contemporary books on the biodiversity and climate crises address healing in mutually supportive ways. Edward O. Wilson's *Half-Earth* is yet another warning of the rapid deterioration of planetary ecosystems. Kathleen Dean Moore's *Great Tide Rising* is a call to action, a plea for "clarity and moral courage in a time of planetary change." These are profound books, simultaneously probing the emotional waves of loss and love. How can you love what you know may soon be lost? Both writers take enormous pleasure in observing, celebrating, and learning from the infinite complexity of ecological and evolutionary systems. The feeling of love is palpable. The necessity of healing is urgent. In their own ways, both writers, a biologist and a philosopher, contemplate and propose approaches for catalyzing that sense of urgency. Nature. Love. Medicine. Healing.

Consider Kathleen Dean Moore's conceptual sequence: "If attentiveness can lead to wonder, and wonder can lead to love, and love can lead to protective action, then maybe being aware of the beautiful complexity of lives on Earth is at least a first step toward saving the great systems our lives depend on." Moore reiterates an important assumption paying attention to the natural world leads to a reciprocal awareness, and the act of response is a form of both personal and planetary healing. Edward O. Wilson suggests that only a shift in moral reasoning will lead to the kind of action he proposes, nothing less than returning half of the earth's surface to wild lands so as to restore the integrity of biological diversity. That's a strong medicinal dose and much healing is required.

But is it enough to love the world? And is it possible to set so much land and water aside? For decades now, ever since Rachel Carson's *The Sense of Wonder*, the holy grail of environmental education has been to cultivate an ethic of care, based on wonder and awareness, spurred by curiosity, and ultimately evoking love. We've pulled out the science (biophilia) to demonstrate how access to landscapes and species promotes the quality of life. We've endorsed regional and national legislative movements ("No child left inside") to mandate outdoor environmental education. We've partnered with organized religion to demonstrate the spiritual synergy of wonder, awe, and appreciation for the sacred biosphere. We've worked with businesses and university campuses to promote sustainable practices. We've advocated for endangered species acts and global carbon reductions. We've explored countless pathways in our efforts to summon action and behaviors that bring ecological awareness to the forefront of consciousness. We've tried to balance the seemingly conflicting impulses of love and vigilance—staying open and wary simultaneously. But it's difficult, perhaps impossible to love the people and institutions who continue to pillage the Earth. It's equally hard to reconcile our own consumer behaviors and actions when they may contribute to declines in biodiversity and increased carbon emissions.

I am suggesting that deepening awareness of the natural world and cultivating love for the biosphere will only take us so far. I know many people who are deeply concerned about biodiversity, climate change, and habitat destruction. Yet they may not be particularly interested in natural history or they may have no background in ecology or environmental studies. I've met great naturalists who don't care one whit about social justice. And I've often wondered how to reconcile Thomas Jefferson's great love of natural history with his slave ownership.

I don't doubt that the way of natural history is virtuous, enlightening, and fulfilling. And I fervently believe that a natural history education, or learning to love the natural world, should be a core learning objective at every stage of educational development. My entire career reflects those beliefs. I've had countless experiences at every stage of my personal and intellectual development that support this view. As a young city child, I felt most free and excited when I was on hikes in new places, when I had access to sea, sky, and open space. My wife and I fell in love in a lovely forest that we described as "the great hall." Our most rewarding family experiences were annual backpacking trips with our children. Two of my books, *Ecological Identity* and *Bringing the Biosphere Home*, emphasize place-based learning and the necessity of intimate awareness of local natural history. And more recently, when I was in the midst of two unexpected surgeries, my recovery was hastened by walks along the shores of Puget Sound, and then a week of hiking in the Tonto National Forest. That was good medicine. When I consider how I want to spend the remainder of my senior years, my priority is to hike more and sit less, sharpen my observational skills, study natural history, roam the hills, explore the world that's right in front of me, and to do all of this while I can. I could happily read about the history of life on earth, continue to learn about ecology and evolution, and use all of these approaches as a means to better understand life and death, meaning and purpose, agency and freedom. This is good medicine now and it will be in the future.

Yet such pursuits, whether I do them alone or in community, as a writer or a teacher, as a grandfather or a wandering stranger, will make only modest contributions at best to the four challenges of ecosystem decline, economic inequity, tribalism, or violence and oppression. Something more is needed. Educators and activists must broaden the constituency for natural history by accepting that it may not ever be a priority for millions of good people who want to lead a better life.

There's another quality, a fifth word that broadens the constituency for environmental education—reciprocity. Nature. Love. Medicine. Healing. Reciprocity. Literally defined, reciprocity is the practice of exchanging things with others for mutual benefit. Reciprocity is a human construction. It implies a cyclical process of altruism, action, reuse, and return. At its most crass materialist level reciprocity implies that we exchange things to derive individual benefit. Indeed, Adam Smith's "invisible hand" is a grandiose, mythical, narrative, metaphorical justification for finding reciprocity in market exchange. On a deeper level, reciprocity assumes benefit that's derived from giving something back, making a contribution for the sheer virtue of doing so. Giving thanks. Lewis Hyde's classic essay, "The Gift Must Always Move" splendidly illustrates the ecological and anthropological origins of reciprocal gift-giving. Every commodity is embedded in a deep personal and community exchange of tribal bonds, ecosystem services, and inter-generational equity. Reciprocity is the opposite of profit. We live in a world in which reciprocity and profit are jumbled and confused, and we often mistake one for the other. A thin line separates mutual benefit from exploitation and it takes a reflective mind to distinguish between the two.

In the contemporary world we equate reciprocity with service, the idea that we devote some aspect of our lives to the welfare of others. Modern colleges and universities often integrate service learning as intrinsic to a good education. We describe service leadership as an approach that puts the good of people and the organization before the specific egotistical needs of the chief executive. We describe service professionals as people who devote their careers to the well-being of others. I like to think of environmental practitioners as service professionals. Most people I've met in the environmental field are service-oriented, and they intrinsically grasp the necessity of service as crucial to their professional and personal success. Service promotes social capital (to stick with the language of profit), and builds community reciprocity, personal satisfaction, and even spiritual benefit. I suggest that loving the world through intimate awareness of natural history must be coordinated with service in community. Otherwise it's merely more self-help, and we have more than enough of that. So I'll add reciprocity to the word sequence. Nature. Love. Medicine. Healing. Reciprocity.

I came of age in the 1960s, and one of the books that inspired my thinking and career was *The Whole Earth Catalog*. Almost fifty years have passed since its initial publication. I often feel as if I am living in a science fiction movie as so many of our predictions from that era have come to pass. In those days, buoyed by our youthful exuberance, we honestly thought that we would change the world. Much has indeed changed, and you can even make the case that there is more environmental awareness today, prompted by the extraordinary global interest in sustainable development. Yet the four challenges of our times remain unsolved. Global sustainability initiatives notwithstanding, the planetary emergency of the sixth mass extinction, plunging declines in biodiversity, changing atmospheric and oceanic circulations, and altered biogeochemical cycles is on our heels. We don't seem to have increased the human capacity to love nature. And we surely haven't increased our capacity to love each other.

Nevertheless, I feel a change coming on. Maybe I've never lost that 1960s naiveté. Maybe I still believe we really can change the world. I have a hunch and I'd like to share it with you. It may be trite to say that communication technology has brought us all closer together. That proximity generates enormous challenges. Some people feel they can keep those problems away by building walls and barriers—the pathological application of the adage "good fences make good neighbors." But many activists, whatever their primary interest, now recognize that all of the four challenges are interconnected. You can't address biodiversity and climate change without understanding the consequences of migration, exile, and the global movement of refugees. You can't meet the challenge of refugees without understanding the necessity of intercultural understanding. Hence biodiversity meets cultural diversity. People concerned with these issues know that those who suffer most will be those in the most deprived economic circumstances. We know, too, that resorting to violence and oppression is a pathological response to the threat of deprivation. It's not the deprived that we fear. It's the prospect that we too may become deprived.

The emerging environmental movement of the early twenty-first century has a new shape and form. Visit any dynamic city and you'll find a new generation at the core of social activism. Communities of color are on the front lines. They aren't asking for more wilderness, or more hiking trails, or wonderful ecotourism resorts. They want opportunity, agency, community health, equal access, participation, and equality. Environmental pollution inevitably is the worst in the poorest, least white neighborhoods. Affordable housing, access to public transportation, access to inexpensive and nutritious food, nearby parks and recreation facilities, the ability to stay put in a gentrifying neighborhood, affordable health care, clean water to drink and bathe in, good schools—these are the environmental challenges for the great majority of people who live on this planet.

This is natural history, too. This is the story of how people live and die, how they earn a living, how they move from place to place, how they mate—the same things we love to observe in the natural world. There's no way we'll preserve the world's ecosystems unless we give equal attention to preserving the everyday lives of ordinary people, or at least acknowledge the close interconnection between these issues.

If you look at the demographics of who reads traditional environmental magazines, let's say *Orion* magazine, or who buys field guides, or who reads *Half-Earth*, or people who take nature walks, you're looking at mainly white baby boomers. And if you look at the demographics of new social and community movements, you're looking at millennials, people of color, and a great variety of groups who in one way or another are dispossessed. The task of educators and activists alike is to redefine the meaning of natural history so it includes all of these groups and all of the diverse sensibilities they entail. Our transcendent challenge is how these diverse stakeholders merge into a coherent, interpenetrating, responsive, and reciprocal social movement. This movement must address the four challenges. The rise of right wing politics reflects a fear of what such a coalition might bring. We will never transcend that divide unless we can look at the root causes of that fear.

Understanding diverse points of view requires generosity, the ability to listen well and reflect carefully about the wishes, dreams, aspirations, and concerns of different communities. We may not always agree, but perhaps we can reach a common understanding and a deeper respect for one another. We can be more generous to those who have different perspectives. This is the sixth and final word of the sequence—generosity. Nature. Love. Medicine. Healing. Reciprocity. Generosity.

I recently visited a groundbreaking, grassroots, community organizing project, coordinated as the Puyallup Watershed

Initiative, located on the southern end of Puget Sound in the state of Washington. The goal of the project is to develop communities of interest (COI), derived from a broad coalition of businesses, nonprofits, government agencies, and interested individuals. These COIs identify the most important regional challenges, eventually coalescing into governance structures that will invest foundation funds into self-generating projects and opportunities. Thus far, the COIs are organized around transportation, agriculture, environmental education, forests, just and healthy food, and industrial storm water. On the ground, this is an inspiring and often difficult process as the players must find ways of linking their specific interests in order to collaborate for the common good. Thus far, the most successful COIs have been those that are willing to coordinate their specific interests, understand the broader spectrum of community issues, and demonstrate multiple forms of generosity, in terms of the time they spend on the project, their willingness to listen to and respect multiple points of view, and their desire to contribute their social capital to the well-being of the watershed.

In essence, the common ground that inspires this project is the widespread belief among all the participants that they live in a special place by virtue of its spectacular geography, its ethnic diversity, and its cultural richness. Ultimately the source of generosity emerges from a love of place, a belief in the people who live there regardless of their background, and a knowledge that the long-term well-being of that place depends on generating respect for all members of the human and more than human communities. It's not enough to love the place. That love must be translated into healing (solving its many problems), reciprocity (deriving mutual benefit), and generosity (giving of yourself to the common good). I don't think it's a stretch to suggest that our word sequence—Nature, Love, Medicine, Healing, Reciprocity, Generosity—is the formula that inspired the original philanthropic gift and that motivates the organizers and participants in the project. Nor is it a stretch to suggest that the project's ultimate success will ride on those qualities.

This isn't all peaches and cream. There are so many struggles and conflicts built into the Puyallup project and others like it. Here are just a few of them. Typically, those who are most generous with their time are those who can afford to do so. That means there are scores of still unheard voices. Some participants have a hard time yielding on the specifics or priorities of their particular issues and abandon the quest for common ground before it even gets started. Some participants lack the experience of working together for a common goal, and don't have the time to get the training that will enable them to practice those skills. Of course there are all of the idiosyncrasies that emerge when people from different backgrounds and orientations have to work together. However, despite these difficulties, the project perseveres, and receives support from a network of like-minded programs that are sprouting around North America. The Collective Impact Forum, for example, is building networks of community organizing principles that support, encourage, and educate about projects like these.

In *Half-Earth*, Edward O. Wilson explains that ecological science is barely in its infancy in terms of what it understands about biodiversity, ecosystem integrity, and environmental change. He refers to unique biosphere habitats, like the edge of the sea, or the surface of water, where contemporary science has yet to discover scores of species that are intrinsic to ecosystem functioning. In other words, we know very little about the very habitats that we are on the verge of destroying.

Surely, Wilson and his scientific natural history colleagues are not capable of "loving" all of the species that they are not yet familiar with. Rather, they cultivate generosity in their willingness to be open-minded as to the great contributions that all of these species make to ecosystem integrity.

Perhaps the same challenges are true of community organizing. How little we really know about social, cultural, and economic organization! And our ignorance stems from our lack of knowledge of ourselves, our psyches, and all of the evolutionary, neurophysiological, and cultural influence on our complex behaviors. Despite all of the wonderful teachings of so many of the world's great spiritual traditions, it's very difficult to love all of your neighbors, whether they are human or more than human. But we can cultivate generosity, open-mindedness, graciousness, and humility in the space of that glorious unknowing. I don't have the capacity to love every species and every person, but I can develop the capacity to be more generous with those people and species that I do encounter.

I believe we won't solve the biodiversity crisis without understanding that ecosystem deterioration, economic inequity, tribalism, and violent oppression are all symptomatic of a deeper malaise. We will not make progress in understanding or resolving any of these issues until we recognize their inextricable interconnections. And environmental awareness alone will not solve the biodiversity crisis. My lack of familiarity with, let's say, the ubiquitous but barely known meiofauna (small benthic invertebrates) is just as consequential as my lack of familiarity with, let's say, the many varieties of Islamic culture. There's not enough time in the day, or years in a life, to become knowledgeable about all of these things, let alone to develop the capacity to love them. Rather I can cultivate a measure of respect for those things I don't understand, and stay open-minded and generous to those who do.

Nature. Love. Medicine. Healing. Reciprocity. Generosity. These capacities, taken together, generate a spirit of community, aspiration, togetherness, virtue, agency, meaning, and purpose. We would be utterly lost without them. The integration of these capacities, taken together, is the source of all of our good works. With the magnitude of the challenges that face us, there is no greater urgency than to understand their significance.

City of Loves

ALISON HAWTHORNE DEMING

Iwalked to the Jardin des Plantes a few days after the terrorist attacks in Paris. I was determined not to let the violence deter me from visiting this treasure during my two-week visit. The city was on alert, armed soldiers patrolling plazas outside the Louvre, Eiffel Tower, and Musée d'Orsay. At the Bataclan makeshift memorials bloomed: long-stemmed white lilies stuck through the slats in a wrought-iron fence, a purple bicycle locked to the fence with a bouquet of white roses exploding through the spokes, a roughly lettered sign "*Je suis musulmune et contre le terrorisme!!!!!*" planted in a field of cheap prayer candles. "I am a Muslim and against terrorism," though I got a laugh misreading, "I am a muscleman and against terrorism." Mounds of flowers still in their plastic wrapping, as if to protect them from the pain in the air, had been laid along the sidewalk, the street populated with the tent city of world news media. Someone had rolled a grand piano onto the street to play "Imagine." Mourners had slipped red carnations through the bullet holes in café windows where days before mortuary vans had lined the street. Soldiers and police patrolled, no idle duty, riot helmets strapped to waists, eyes laser sharp scanning the crowd, rifles cradled, fingers on triggers, muzzles pointed downward but poised to lift. Brussels had closed its metro in fear of chemical or biological weapons. The Paris metro remained open, though it was easy to feel alarm riding through those ghostly tunnels—especially one evening when I rode through stations where not a soul was in sight. Did others know something I did not? But Paris was defiant. The icon of the Eiffel Tower morphed into a peace sign and popped up in café windows with signs reading "Not Afraid." People's eyes grew sharp with fear, yet Parisian friends said they only thought about the attacks if they turned on the news. "Otherwise, I go to work. I go to bars."

When our host at a dinner party served a gorgeous home-made apple tart, all the apples lined up like overlapping quarter moons, someone joked, "This is what they want to kill us for." Our hotel host smiled and asked, "Your visit has been good? How did you like our city of abomination and perversity? That's what they called us." "Yes, we still love Paris." "Good, Paris needs your love." She said the terrorists were trying to start a civil war but it wouldn't work. She posted a photo of herself with her Muslim friend, the two encircled by a heart. She was in Bayonne visiting her mother when the attacks came. I need to come back, she told herself. Her mother said, Stay. No, she said, Paris is home. I need to be there. It will not work, she said, what they are trying to do.

At the Jardin des Plantes, I walked past the statue of Buffon at the entrance, the bronze sculpture made in 1908 in which he is draped in a dignified coat so grandiose it threatens to swallow him, his hair coiled into three horizontal pipes on either side of his head, a bird in his hand about to take flight. It is the visage of a master, as Buffon, the father of natural history in France, surely was. Born George-Louis Leclerc in 1707 into a family of civil servants, he became an aristocrat thanks to an inheritance with which his father purchased an estate that included the village of Buffon. As a young man he went on a year-and-a-half grand tour with his pal the English Duke of Kensington. During that time he gave himself the title Comte de Buffon. What makes him memorable is not that achievement of class, but his revolutionary work in science that led him to challenge two thousand years of dogma about the origin of species and the age of Earth. One hundred years before Darwin, Buffon penned his forty-four-volume *Histoire Naturelle*, works read as widely as those of his contemporary Voltaire. He studied first with Jesuits, then studied law, then mathematics and medicine. He made an extended study of the properties of wood used in ship construction. He patiently tested more than one-thousand small specimens of wood, then realized he could not learn what he needed to know without studying large pieces and so realigned his study.

In 1739, Buffon was appointed Keeper of the Jardin du Roi, his job to catalog the royal collections of medicinal plants. The garden had been opened to the public in 1645. Buffon's aspirations far exceeded his job description. He aimed to catalog the whole natural world, studying minerals and animals and plants. He grew the garden into a major natural history museum for which he gathered specimens of flora and fauna from around the world.

Buffon disagreed with church dogma that Earth was six thousand years old. He thought it much older. He disagreed with Linnaeus's idea of classifying species with clear delineations. He saw gradations between species and individuals that made the definitions less clear. He found it curious that similar environments in different regions spawned different plants and animals—the idea behind the science of biogeography. He believed that life, like Earth, had a history—the idea behind the science of evolution that would advance over the next century. So prized was his work that his cerebellum is sequestered in the base of another statue of the naturalist, this one in marble, made in 1776 by the sculptor Augustin Pajou, depicting a younger man, standing beside a globe, lion and snake at his feet, dressed in a toga from the waist down with one bare leg exposed up to the thigh, folds of cloth draping down over the globe as if he were unveiling its secrets, his chest bare and virile, his locks long, curly and spilling onto his shoulders.

In the Jardin des Plantes, I walked past an old woman doing tai chi beside a garden of medicinal plants, past the giant Cedar of Lebanon and down walkways lined with bare-limbed plane trees of autumn scratching the sky. I marveled at the beauty that can be built with living plants given a few hundred years of tending. I lingered at the Dodo Manège where children circled round and round riding on the backs of dodo, tortoise, panda, gorilla, glyptodon and Tasmanian wolf—extant and extinct brought together and animated for the delight of children. I walked past Aristotle, his stony surface warmed with a gray patina draping over his bare shoulders. Seated, he was contemplating the egg he held in his left hand that was resting on his left thigh. Which came first, the chicken or egg? he asked. He was lost in thought, and the little egg had become almost too heavy to hold. Scrolls of completed manuscripts were rolled up and tucked behind his chair. No matter how much wisdom he accrued, he would be forever lost in wonder to the mystery of existence, the only answer his own inwardness and silence. How perfect that he could hold that pose there beneath the inwardness and silence of the arboreal ancients.

In the Jardin des Plantes, I approached the Grand Gallery of Evolution, stepping up the worn stairs to encounter a spectacle of ghostly animal force. The stampede of pristine skeletons galloped head-on toward me, horses and zebras with leading hoof raised and head leaning forward in flight, gaters and crocs floating in their glass case of air with jaws wide, herons and cranes, sturgeons and hares, cats and catfish, all posed and poised as dancers in a macabre celebration of fixed movement. Hearts and inner ears floated in formalin vases. You must take life all apart to understand it, the relics said. Here is the lining of a giraffe's esophagus, the terrain of the tissue bearing the same irregular pattern of the plaid on its pelt. Here are a set of brains attached to their stalks and spines, standing upright in narrow vases with a blue glass backing to accentuate detail. Here we are all just lilies with brains above ground and our feet wading in the sea. And the brains, shocking, like so much tubing packed into the skull. Here is the sine wave of a python skeleton suspended in space, here the arteries of a human brain teased out and spread like a flower—everywhere deltas and tributaries, ripples and waves. Hearts and their piping—the masters of flow.

Even the monsters are represented—cyclops cat, faceless fish, pig fetus with one head and two bodies, two lambs joined at the chest. And the thing is that everyone looks happy here among the dead in the Père Lachaise of the world's fauna. The Grand Gallery of Evolution exists to serve science, to catalog and detail and define our fellow creatures and to learn how we all got to be the creatures we are. Yet the visitors to this place are not making scientific illustrations or gathering data. They are staring in astonishment, telling one another, Look at this! Did you see this?! They are feasting on the wonder of life, filling themselves on nutrients they did not realize they lacked. They are like the deer who come out of the forest in winter to lick salt off the beach rocks at low tide then gambol in the waves. Here is the heron's neck, bones like the links on a bicycle chain. People are being reunited with their distant relatives, all these beings that have lived a life of movement, appetite, decisions and daydreams and, like us, improvised their way through the forests and plains and waters and deserts where predators might leap but most days they did not.

I had not come to Paris to learn about natural history, nor had I come to learn about terrorism. I came in search of family history, my great-grandmother, now long among the buried bones, who had been born in Paris and became a dressmaker for Empress Eugénie during the reign of Napoleon III. But terrorism sent me on an errand into a new wilderness, one where the opulence of the nineteenth century court, the theater of aristocracy that artisans such as my great-grandmother worked to create, faded into the startled apprehension of the present moment. I became animal in those days, tense to the environment, knowing someone/something might kill me just because I was alive. Was that a suicide vest bulking up the jacket of the man in line ahead of me at the Louvre? Ah, no, only a copy of Apollinaire's poems shielded from the light rain drizzling upon the gray cobblestones.

It is easy to claim the healing power of nature. Just look at the medicinal gardens that have offered their testimony for centuries: Foxglove gave us digitalis for the heart, Aloe Vera gave balm for burns, willow gave aspirin for pain and the ginger family gave turmeric for inflammation. And it is easy to claim that I was healed from the flood of anxiety that the attacks had produced in me by walking in the Jardin des Plantes. I did not need to partake of its salves and potions but simply to be in the presence of that orderly care and cultivation as antidote to the annihilating forces that lurked. The garden was a statement of the virtue of tending, and the mark of care on those plots became an act of care inside me. It is not easy to claim that the cult of death that needs to know nothing about its victims in order to want them dead can be healed by the power of nature. The tactical viciousness of the Paris killers, one of whom entered a café, sat down and waited until he had placed his order to detonate his suicide vest, suggests a businesslike calm to the enterprise, a perversion of the spiritual precept of being present in the moment. Monsters. "Gardens," writes Robert Pogue Harrison, "become places of rehumanization in the midst of, or in spite of, the forces of darkness."

Paris is known as the City of Love, the place for lovers to stroll along the dreamy banks of the Seine and find in their appetite for each other an escape from the world that is, as Wordsworth wrote, "too much with us." Denise Levertov refuted the claim when she wrote, "the world is not with us enough," and she insisted on poetry that brought our desires out of the bedroom and into the sphere of public life. I remember when she came to Tucson to give a poetry reading sometime in the early 1990s. I took her for a drive to see the desert. She was aghast at its expansiveness. But it was the vast cotton fields that made her want to stop the car and look more carefully. The fields had been harvested and this seemed a marvel to both of us that such arid land could be cajoled—and water arrested from its natural flow to make it happen— into producing miles of cotton fields. The cotton bolls, still cupped in the rigid spreading dried pods, had been baled into giant bricks the size of a semi-trailer, the fluff of cotton spilling out from the prickly dross. She gathered one of the pods, a weedy-looking nothing, and held it in her hand, stroking the little puff of cotton as we rolled along the Interstate. It was a moment that has stayed with me, the opportunity to witness the poet's wonder, curiosity, and receptivity to experience. Cotton, which now signifies for many people a healthy and natural fabric but which two-hundred years ago in America signified the evil of slavery and the moral filth of the money it helped our nation to amass. Cotton. Just a plant. But also the totem of our burdens and hopes.

Wonder, curiosity, and receptivity to experience: these are the attributes of the poet and naturalist. Animal attentiveness and humility in the presence of the Grand Gallery of Evolution: these are the attributes of the acolyte of life who stands on the knife-edge where the hiker tests her courage against the dangers that fall away to either side of her passage or where the rider on the metro agrees to the contract of city life: I embrace you, stranger. May we live in love.

The Path of Healing

JUDITH LYDEAMORE

The day started like any other. I kissed my partner, John, goodbye in the morning, said goodbye to our dogs, drove to work, and settled down to the day's tasks. Four hours later I was fighting for my life.

Lucky for me, I was in the middle of a phone conversation with my manager when my brain hemorrhaged. I was gripped by grinding pain across the top of my head and behind my right eye—the worst pain I'd ever experienced. Also lucky for me, my manager heard me drop the phone and immediately called for help.

I don't remember my professional assistant coming into the room but I do have very vivid memories of what happened next. Jenny said she'd found me standing, clutching my head with one hand and holding the edge of a table with the other. She asked how I was feeling, and should she call an ambulance. "Yes!" I'd never felt such pain before.

One of my team called John. He said that at 11:20 everything was going along in an ordinary way, and at 11:21 everything changed. He flew out from his work and drove to my office. I remember hearing John's voice and feeling safe.

Next I was in the Emergency Department of the Royal

Adelaide Hospital. "What's your pain level like?" one of the staff members asked. "Eleven out of ten," I replied.

A little later John asked me how I felt and I remember saying, quite calmly, "I think I'm going to die." With that, John stepped up to the emergency desk and said, "I don't think you're taking this seriously enough. Judith's just said she thinks she's going to die. She wouldn't say that if she didn't mean it."

Real action began immediately.

I was taken for a CT scan and then John was given the diagnosis: an aneurysm had ruptured, resulting in a level four subarachnoid brain hemorrhage. Some people experience excruciating headaches which alert them to the situation. Not me. In fact one of my catch phrases used to be, "I seldom get headaches." I don't say that anymore.

"It doesn't get more serious than this. A third of those who suffer a brain hemorrhage die immediately, a third will die in hospital or be left with severe side effects, and a third will recover. She's over the first hurdle."

That evening I underwent a five-and-a-half-hour craniotomy—a team of doctors cutting a large piece out of my skull to access my brain, clip the aneurysm, and then clip my skull together again.

I have very few memories of anything that happened during the next week, but I kept remembering a tree of light—a tree with branches and twigs everywhere, and all of it shining in exquisite detail, illuminated by a brilliant white light. It was only much later that I worked out what I had "seen."

My brain was in crisis, inflamed by the trauma and the blood seeping out of the burst aneurysm. To take the CT scan, dye was pumped into my blood vessels and the "tree" was every artery and vein in my inflamed brain lighting up with irradiated dye.

I slowly began to regain my self-awareness and also to wonder and worry about what had happened. What day was it? What had been going on? John found a practical solution. He wrote down days and dates and events from the time of the hemorrhage. I clung to that list and used it to help reorient myself to my very changed circumstances.

The physiotherapists had me up and walking during my second week in hospital, doing small exercises to help determine what damage might have been caused by the hemorrhage.

One day they suggested I go for a "walk" in a wheelchair outside the hospital. The old Royal Adelaide Hospital is next to the Adelaide Botanic Gardens. Whether this juxtaposition was deliberately planned by the early colonists or whether it was a case of serendipity, it meant a place of nature next to a place of healing. John decided he'd take me into the Gardens.

I must have looked a sight! Head bandaged, in my nightgown and dressing gown, being wheeled along. I didn't think of that. I was feeling the warmth of the sun seeping into me, listening to the traffic and people talking, and all the day-to-day ordinariness of life going on.

But also feeling every little bump in the footpath jolt my head, which I held to stop it from hurting.

We turned into the Gardens and it was instantly much quieter—the embracing quiet you feel when you step into a forest, and yet twanging with energy. I remember saying to John, "It's so green!" It was calm, peaceful and serene, with the city sounds muffled. And the color contrast between the white walls inside the hospital and the green everywhere in the Gardens was overwhelming.

Suddenly, as if on cue, along came some beautiful

Australian Wood Ducks and their ducklings, walking and talking. Ducks! I had to smile. Our wood ducks look a bit like a small goose and are comfortable around people. They constantly chatter to each other, as well as stop to look and consider what is going on around them.

The Botanic Gardens are an oasis in the heart of Adelaide, a tranquil space, and a haven for birds and insects. Green trees and grass, bird calls, wood ducks, sunshine and warmth.

It was a hot summer day and the air was filled with the sound of insects buzzing and humming. I hadn't heard that constant and familiar background noise to a summer's day for a while. It was so very different from hospital sounds.

Birds were all around us: rosellas calling, Rainbow Lorikeets screeching, wattlebirds uttering their raucous calls, ducks chattering. It was a symphony of noise that lifted my spirits immeasurably. Australia is a continent of birds, and the Botanic Gardens are home to many city-dwelling birds. Every now and then we stopped and looked and listened, inhaling the air. The smell of eucalyptus exuding from the gum trees in the heat was everywhere.

Nature's delights in abundance and freely given.

The path we were on led to the Botanic Gardens bookshop. Now I know it was also the path leading to my healing. Books and words have been part of my life forever, and books about Australian plants and birds and animals are favorites. We just had to go in. Bless the woman who was serving that day. She took my bandaged head and dressing gown all in her stride with a cheerful smile as if to say, I know how important this is. Which it was—I was back on familiar ground.

And then my joy of being with books re-started. I couldn't stand up to reach the books I spotted so John, bless him, picked up the ones I pointed out and handed them to me. "Oh, I have to have that . . . and I have to have that one . . ." John later said he knew I already had a few of them but I was enjoying myself so much and I was so alive with interest that he just said, "We'll have that one too." Lots of books, and a reconnection to the things that are important to me. I tired quickly but was filled with the sights and sounds and smells and warmth of the day and of life outside the hospital's walls.

John got me a television with access to a satellite network and I asked for the natural history channel. He brought in subscription magazines that had arrived since my brain hemorrhaged. They came from the Nature Conservation Foundation, Birds Australia, and Native Grass Resources Group, to name but a few. I do believe my full recovery began when I reconnected with nature, and with words and natural history, or natural history in words.

Going-home day finally arrived. John said we needed to buy a few things in Stirling, our local town center. We parked along Druids Avenue, which is lined with oak trees planted in 1890 by The Pride of the Hills Lodge of the Druid Society. It was December, at the height of our austral summer, and these 120 year old oaks were in full leaf. The trees were very green, and bright light filtered down through the leaves. I enjoyed waiting while John did the shopping. I saw people I knew, cars were going up and down the road, children and dogs were running around. It was so normal and everyday.

Finally we headed for home. Everything was so familiar and yet so different—as if I had been born anew after my near-death experience.

Our four-year-old Labrador retrievers, Sophie and Pippa, met us at our gate. John got out and said, "Mum's home." Instead of leaping up to greet me, as they usually did after I'd been away, they came up to the car door, gently sniffed me to say hello, and very quietly stood back so John could help me out. From that moment on, their company was a priceless part of my healing.

We share our world and our lives with myriad other living beings in nature, and the thoughtfulness with which our animal compatriots assess what's going on and respond appropriately astounds me each and every day.

Driving down to our house I saw a native orchid. "Look at that! I don't remember seeing that one before. Will you take a photo please? I'm not sure what it is." John got me safely inside and into bed then went up and took some photos. It turned out to be a pink hyacinth orchid, *Dipodium roseum*, with a brown leafless stem and pink flowers with spots all over the petals. Focusing on these minute botanical details brought me out of my own worries and into the larger order of the world.

Outside our bedroom window is a bird bath. It was a very hot summer so lots of our beautiful native bush birds came in for a drink and a bath throughout the day. They kept me company over my many months of recovery, as did my family and friends who stayed nearby to ensure nothing went wrong. I added many new sightings to the home bird list I'd been keeping for eight years, including an extremely rare bird for our area, a Black-chinned Honeyeater.

Three months into my healing, we had to replace our solar hot water service. The water tank on the roof needed to be drained so the repairs could take place. In our summers water cannot be wasted. What falls from the sky is what we have for use throughout the year. I decided to sit outside with a range of collection drums and catch the water to use in the bird baths and on the potted plants we enjoy. I sat at the back of our house, still not very well but with bins at the ready, and the water was released.

I shifted the hose from one drum to another as each filled, and then chattering parrots appeared in the trees in front of me, feeding. Their raucous calls drew my attention. Mind you, I didn't know at the time which parrots they were, so all my bird watching skills came to the fore as I registered the distinguishing features to look up later in a bird book—red forehead, blue crown, yellow band on the wing—and yes, they were definitely Musk Lorikeets, a new entry for our bird list.

As I reflect back now, I realize how important it was for these lorikeets to appear, for me to take note of them, look them up in our field guides, and add them to our list. I was reassured that my mind was working well. I was paying attention to the world around me, picking up details of sights and sounds, just as I used to before the brain hemorrhage.

I was well on the path to recovery and healing.

Biophilia at My Bedside

ELISABETH TOVA BAILEY

From my hotel window I look over the deep glacial lake to the foothills and the Alps beyond. Twilight fades the hills into the mountains; then all is lost to the dark. After breakfast, I wander the cobbled village streets. Huge bushes of rosemary bask fragrantly in the sun. I take a trail that meanders up the steep, wild hills past flocks of sheep. High on an outcrop, I lunch on bread and cheese. Late in the afternoon along the shore, I find ancient pieces of pottery, their edges smoothed by waves and time. I hear that a virulent flu is sweeping this small town.

A few days pass and then comes a delirious night. My dreams are disturbed by the comings and goings of ferries. Passengers call into the dark, startling me awake. Each time I fall back into sleep, the lake's watery sound pulls at me. Something is wrong with my body. Nothing feels right.

In the morning I am weak and can't think. Some of my muscles don't work. Time becomes strange. The streets go in too many directions; I have difficulty finding my hotel. The days drift past in confusion. I pack my suitcase, but for some reason it's impossible to lift. It seems to be stuck to the floor. Somehow I get to the airport. Seated next to me on the trans-atlantic flight is a sick surgeon; he sneezes and coughs continually. My rare, much-needed vacation has not gone as planned. At this point, I just want to get home. After a connection in Boston, I land at my small New England airport near midnight. In the parking lot, as I bend over to dig my car out of the snow, the shovel turns into a crutch that I use to push myself upright. I don't know how I make the long drive home. Arising the next morning, I immediately faint to the floor. Ten days of fever with a pounding headache. Emergency room visits. Lab tests. I am sicker than I have ever been. Childhood pneumonia, college mononucleosis—those were nothing compared to this.

A few weeks later, resting on the couch, I spiral into a deep darkness, falling farther and farther away until I am impossibly distant. I cannot come back up; I cannot reach my body. Distant sound of an ambulance siren. Distant sound of doctors talking. My eyelids heavy as boulders. I try to open them to a slit, just for a few seconds, but they close against my will. All I can do is breathe.

The doctors will know how to fix me. They will stop this. I keep breathing. What if my breath stops? I need to sleep, but I am afraid to sleep. I try to watch over myself; if I go to sleep, I might never wake up again.

🌿 🌿 🌿

At age thirty-four, a mysterious pathogen zapped my mitochondrial function and sent my autonomic nervous system haywire. It was frustrating enough to suddenly be ill with a disease that the medical world couldn't cure, but there were the additional challenges of isolation from society and the natural world. I longed to hike the forest trails that surrounded my house, to follow the river's path that edged my property, and to be out in the seasonal weather shifts that I could only glimpse from a far window. I was too weak to hold a book, too ill to sit up to watch a film; even quiet music wore me out. Though I loved visits from friends, even those exhausted me.

During one of my bedridden years, a terrarium made by a relative became a welcome oasis for my mind. Its small green world distracted me from the intolerable symptoms of illness and the myriad worries of my disabled life. Everything in my bedroom was completely still, inanimate; but in the terrarium, life flourished.

The terrarium's floor had a layer of upside down sphagnum moss. This was covered with a few inches of forest loam, which surely contained some seeds and insect eggs—a fertile mix of surprises. Nudged into place on top were mosses, polypody ferns, partridgeberry, wintergreen, woodland violets, and goldthread. At first, the plants looked uncertain in their new home, but in a few days they began to reach for the sun as it angled through the window, and they worked out elbow room with their neighbors. Then everything began to grow, and soon this patch of forest floor, with its rich landscape of green textures and shapes, looked as if it had always been there.

Where there is life, there is plot, and in this green kingdom, just inches from my bed, the lives of tiny creatures were unfolding. One night I woke in the dark, turned on the light, and saw a juvenile slug making its way along the terrarium's glass wall, high above the plants. Until that moment of discovery, its nocturnal life had secretly dovetailed with my diurnal one. An ant would appear out of nowhere on a run, and vanish just as quickly. What was happening in the terrarium that caused its frantic rush? Or was the ant incapable of moving at a slower speed? A minute spider wove its web strategically at the edge of a fern stem and a dip in the moss, like a clever, if sinister, newsstand proprietor at a subway entrance. A mosquito hawk hatched and flew around, occasionally, a moth fluttered by; both would be freed in the spring. The terrarium's ecosystem was evidence of life on a small scale; it contained the inhabitants that I knew—the spider, the mosquito hawk, the slug, the ant—and the inhabitants they knew well that I didn't—the minuscule creatures too tiny for me to see.

My terrarium held another particularly interesting nocturnal animal. A white-lipped forest snail, *Neohelix albolabris*. A friend out for a walk in the woods had picked it up and brought it along when she came to visit. The snail took up residence first in a pot of violets by my bed and then in the terrarium. I spent a year observing my gastropod companion and getting to know its habits and preferences. My room was so quiet that I found, to my surprise, I could even hear the sound of the snail eating. Eventually, I would read about the natural history of the gastropod and learn of my snail's thousands of teeth. I read about the complex biochemistry of its slime making and the pheromones it left in its trails. The intimate details of a snail's life were common knowledge in nineteenth-century Europe, where snails were farmed. Eventually, through the lyrical writings of the Victorian naturalists, I would learn of the passionate love life of the gastropod.

As my illness excluded me from most of my own human world, the snail's world became my world too. The snail was beautiful to watch as it glided along, and its slow pace calmed my mind. Its adventures among the varied green mosses were a welcome contrast to the stark white walls of my bedroom.

Looking into the terrarium, I also observed the day-to-day unfurling of a fern frond. At the time, I did not know of Dr. Nathaniel Bagshaw Ward, a nineteenth-century physician with a passion for botany, but I have reason now to be grateful to him. Dr. Ward had tried to grow ferns outdoors in London, but no plant could survive the city's coal-smoke pollution. He also had a passion for entomology, and in 1829 he placed a chrysalis and some leaf mold in a bottle. He topped the bottle with a piece of tin and put it outside his window. As he awaited the chrysalis's hatching, he was amazed to see a fern begin to grow and even thrive in the bottle's clean ecosystem. He had accidentally invented the terrarium. He named his concept of glass containers for plants "closed" or "Wardian" cases and they quickly became a Victorian fad. Everyone in London, from working class to wealthy, had to have one. They ranged from simple, inexpensive designs to elaborate palace-like constructions.

Dr. Ward's interdisciplinary interests led to another innovation, one that I was now discovering on my own: the use of a terrarium to improve a patient's environment. During the nineteenth century there were many medical misconceptions, including the fear that indoor potted plants stole oxygen from the air and would be dangerous in a sick room. However, the enclosed space of a terrarium solved this problem and Dr. Ward noted its benefits in his book, *On the Growth of Plants in Closely Glazed Cases*:

> As a means of administering comfort to the afflicted and distressed in body or mind, [terrariums] are invaluable. . . . [whether] confined, from paralysis or other causes, to a bed or sofa, either in country or town, [patients] have thus been enabled to beguile many a weary hour. . .

Just as Dr. Ward would have predicted, I spent hours each day looking into my bedside terrarium, yet I mentioned its existence in my life to only one of my own doctors; most of my medical professionals spoke only the language of tests and lab results. Despite more than a century of improvements and innovations in the medical field since the Victorian era, the experience of being ill remains the same: discomfort, weakness, pain, uncertainty, and an inability to take part in normal life. Illness too often vanquishes the meaning of life.

In 1842, an editorial in the *London Quarterly Review* by an unknown writer described an all-too-familiar scene:

> Who is there that has not some friend or other confined by chronic disease ... to a single chamber, one ... who, a short while ago, was among the gayest and most admired of a large and happy circle, now, through sickness, dependent ... for her minor comforts and amusements on the angel visits of a few kind friends.

When a chronic illness severely reduces a patient's world, how *does* one survive? Illness can too easily create a prison from which it is difficult to escape: a prison of the body, the mind, the room in which one exists. The unknown writer continues:

> In the evening, a largish box arrives [from her physician. It contains] a large oval bell-glass fixed down to a stand of ebony, some moist sand at the bottom, and here and there, over the whole surface, some tiny ferns are just pushing their curious little fronds into life ... Every day witnessing some change, keeps the mind continually interested. ... The doctor, the next morning finds the wonted cheerful smile restored, and though recovery may be beyond the skill, as it is beyond the ken of man, he at least has the satisfaction of knowing that he has lightened a heart in affliction, and gained the gratitude of a humble spirit.

Imagine a physician so keenly aware of the humble limitations of medicine and his patients' plights that terrariums are a routine part of his treatment regime. Dr. Ward knew that a view into a terrarium could elicit hope and provide a way to stay connected to the world. His insight into the importance of patient environment far surpasses that of our own contemporary medical world. Dr. Ward's son Stephen, also a physician, addressed the English Royal Institution in a lecture titled "On Wardian Cases for Plants, and Their Applications":

As a medical man, I have long felt deeply interested [in] the aesthetics of the sick-room...and in the treatment of those of a chronic nature, where the cure is tedious...A sum of money has been collected ...for the construction of [Wardian] cases to be placed in the wards of the [London Hospital]...I feel assured that their presence will be both gladdening and beneficial to the patients.

I like to imagine the London Hospital rooms of the 1800s graced with fern-filled terrariums. Biophilia—the critical connection between human beings and the natural world— should be a key part of patient care. Medical facilities should be designed to give patients close window views of the natural world and why not fill our hospitals and assisted living centers with terrariums? Healthcare professionals could query patients on their home environments and encourage a connection to the natural world.

Terrariums have gotten me through the coldest winters and the worst stretches of illness; they hold the promise of spring and the hope of convalescence. When the snow deepens around my house, I watch fern fronds stretch upward then delicately arch over, forest plants send out new runners, and green mosses grow lush. A terrarium is a habitat for the unexpected, both real and imagined. It's a microcosm that contains all of life from birth to death: the challenges, intricacies, and mysteries.

Murtle Lake Rx:
A Dose of Wind and Rain

SAUL WEISBERG

Day 1 – *Low clouds, light sprinkles of rain, the air is heavy and still.*

The portage into Murtle Lake is just over a mile and a half, an acceptable sacrifice to access the largest paddle-only (non-motorized) lake in North America. The bugs, while fierce, are manageable, as my wife, Shelley, and I push and pull our canoe cart over a winding trail following a small creek toward the water. At the lagoon, we pack two weeks' worth of food and gear into our canoe and begin to paddle as thunderclouds form on the horizon and the wind begins to rise.

Murtle Lake is a large wilderness lake nestled at 3,500 feet in Wells Gray Provincial Park in east-central British Columbia. The Y-shaped lake is divided by Central Mountain—the West Arm encompasses ten miles of small coves, large lagoons, and small islands, while the North Arm extends twelve miles into the Cariboo Mountains where the glaciated Wavy Range dominates the skyline.

Glad to be on the water, we paddle along the meandering shoreline, skirting small marshes and scattered sandy beaches, and passing over black outcrops of volcanic rock. Wind and sun are interrupted by short periods of rain.

As we move deeper into the West Arm, we surprise a female Barrow's Goldeneye and her recently hatched brood of ducklings and follow them as they appear and disappear in the scattered reflections of sunlight on the water. Three would fit into the palm of my hand. A couple of coves later we come upon a female Common Merganser with seven young. They hear our canoe and launch into a splashing near-flight around a corner and out of sight.

Shelley's laughter mixes with the call of a Common Loon from somewhere down the lake and we settle a bit deeper in. Encounters with birds are often one of the first things that remind us we're in a new place, that we're finally *here*. All the packing, the long drive north, work, the unanswered emails fade away. Just two of us, a canoe, a new lake, and two weeks stretch ahead.

Day 2 – *Morning sky clear, a few small clouds, light breeze from the west.*

We wake early after a night listening to the gentle splashing of waves. Our simple camp sits on a thin stretch of beach buttressed by dense forest claimed by a pair of Spotted Sandpipers, and hundreds of small, dark toads hopping through the vegetation at the edge of the woods.

John Muir mused: "I only went out for a walk and finally concluded to stay out till sundown, for going out, I found, was really going in." I read that as a teenager and it's stuck with me ever since. *Going out*—to connect deeply with wild nature—is a choice I make several times a year. It pulls me away from the comforts of home and the responsibilities of work. I need that deep immersion in nature to be healthy, to feel complete.

As a child I lost (and found) myself by going outside.

Tents and tarps provided cheap vacations for a family without much money. One family camping trip led to another and national parks and forests became old friends every summer. In my teens, I discovered birdwatching, which led me to natural history and, later, field biology. Nature was the place where I found the space and time to watch and listen and *feel*. It still shapes how I engage with the world. I discovered the simple tools—binoculars, field guides, daypack, ice axe, paddles, maps—that led to new adventures. To this day, a small notebook for poems and field notes is always within reach. It helps me remember the details and make sense of the stories that emerge from wild places.

After breakfast, we pack slowly, taking time to find and sort gear and balance the weight of packs, fore and aft, side to side, in the boat. We leave camp in midmorning and paddle deeper into the West Arm as an intermittent rain begins to fall. The sky darkens before we find our next camp, hidden behind a small beach along a forested shoreline. As we ferry gear to our campsite the rain becomes stronger and thunder sends us scurrying to set up tarp and tent. Often you don't have much choice in canoe camps—you set up where the topography allows. This camp is small and beautiful, the ground covered with bunchberry and twinflower glistening with raindrops.

Several large *Populus* trees at the water's edge give this camp its name—Cottonwood. Our site was carved from the surrounding forest, a dense mix of Western Red Cedar, Douglas-Fir, Western Hemlock, Engelmann Spruce and a few Lodgepole Pine. The Mountain Pine Beetle population explosion in the last ten years has changed these northern forests, and dead pines stretch gray limbs toward the sky. In some of the openings, huckleberries and ferns fill the gaps with green.

Later in the evening, the air is still, the afternoon thunderstorm a memory. We sit in camp under clearing skies and listen to the soft calls of three loons who have entered the cove below our tent. The far shore of the lake echoes their cry.

Day 3 – *Heavy rain last night. We wake to wind and clouds. The lake is a gray sheet of waves moving northward.*

It's a good day to climb a mountain. A half-hour paddle to the trailhead where we follow a brushy trail uphill listening to thunderstorms advance across the valley. Several hours later we stop at a level spot in a grove of Silver Fir, Engelmann Spruce, and Mountain Rhododendron and sit on the ground snacking on rice crackers and smoked salmon. Shelley has brought chocolate. Thunder turns to rain which turns to hail. It feels like the light is being sucked back into the sky.

It's good to remember I can be comfortable and happy hiking uphill on a steep trail through thick brush in a rainstorm. The wet brings a special quality to the light and wildflowers glow along the trail. Back at camp we hang our soaking clothes from the centerline of the tarp and make hot drinks, looking out at the wind-tossed lake.

In 1901, in one of the first books on national parks, Muir further developed his thoughts: "Thousands of tired, nerveshaken, over-civilized people are beginning to find out that going to the mountains is going home; that wilderness is a necessity; and that mountain parks and reservations are useful not only as fountains of timber and irrigating rivers, but as *fountains of life.*"

Muir was right. When I walk into a forest or paddle on a lake my body and mind connect in a way that feels healthy and complete. Muir's "fountains of life" feed my body and heal my spirit. When needed, they comfort my soul. I fall in love with the world all over again. And while many people share this experience, and seek it out on a regular basis, many more don't. Sometimes I'm okay with that, especially when I look out across the lake and relish our solitude. But I worry about the future: without direct experience of nature, how will people develop the political will to protect these landscapes? As I sit on this wild beach, looking at the windswept lake from underneath our dripping tarp, I am a participant, not merely a spectator. The wind chills my bones. My gloved fingers hold my field notebook whose pages are covered with wet graphite smears. The raven's croak from the spruce behind our camp is part of me.

Day 4 – *Cool morning, low overcast, winds from the west, mosquitoes a constant presence.*

I spend most of my life ruled by the tyranny of clocks. Wild places reveal new ways of connecting, or disconnecting, with time. On the lake I get up with the sun. I paddle west. I walk uphill, swat mosquitoes, look closely at moose scat, and write in my journal. Shelley teaches me the names of new flowers. I remember how much time there really is in a day and a night as we re-learn the natural rhythms of nature.

I feel tired, happy, *healthy.*

The evidence is getting stronger every day that spending time in nature is really good for people. I've been thinking about the health benefits of nature a lot lately. It connects to my work as director of North Cascades Institute, a conservation organization whose mission is *to inspire and empower environmental stewardship through transformative educational experiences in nature.* It's my job to figure out how to do it. In essence, I've been challenged to ask: *what is an effective dose of nature?*

I laughed when I first heard this question. It seemed silly. I don't think *too much nature* is a problem. I've joked about being close to an overdose several times, but things usually looked better in the morning when the storm passed. I recognize that some people appear addicted to nature, but I consider this more of a healthy co-dependency. And while withdrawal is unpleasant, it's easily remedied by the next trip to a mountain meadow. We need more data—the same as we've compiled on nutrition, sleep, and exercise—on how nature might change our behavior and our health—as individuals, organizations, and communities.

Day 5 – *Rained hard all night, continuing this morning. Woke at 6 AM to wind and waves, crawled back into my sleeping bag.*

A bumblebee flies into camp, circles each five-gallon food bucket, inspects my tea, flies away into the mist. We break camp and are on the water by noon, heading north. A good day for paddling—a lively mix of tailwind, headwind and bright sun mixed with cold rain—eight miles in two and a half hours. We camp halfway up the North Arm at a small beach with a spectacular view of the snow covered Wavy Range rising above the lake's eastern shore. The afternoon drizzle turns to evening sunshine. Shelley's voice breaks the silence: "Rainbow!" A brilliant streak of color rises from the lake, arcs over the mountains, curves at the peak, and drops to the lake a few miles to the south. The colors darken and glow. A second rainbow appears, somewhat fainter. The lake becomes glass and the bright reflections darken and race across the water. In the quiet after dinner, Shelley grabs her binoculars: the far shore is roiled by breaking waves and whitecaps. They stream from the north, move toward the middle of the lake and within minutes the entire basin is a mass of froth and foam. We sit amazed, grateful for being here, sharing this moment, our snug camp, and mugs of hot tea.

Day 6 – *Rain last night. Cold wind this morning mixed with more rain. Low, overcast, gray. High clouds move in from the south.*

The beach at our camp is about one hundred feet wide by twenty-five feet deep. There is no other place to walk except for a serious uphill bushwhack through a wet dark forest. I sit, then pace back and forth, watching loons and a Common Merganser with six chicks that scramble to ride on her back. Only three can find a holdfast as they alternate climbing on and falling off into the rippled waters.

Standing at the water's edge, I think about the research that demonstrates clear links between exposure to nature and physical and mental health, social well-being, and behavior: exercise in green environments reduces depression and improves self-esteem; settings that include trees and open space reduce stress; nature walks can improve children's attention; residents of public housing report stronger social bonds and less violence when there are trees near their apartments; and in neighborhoods close to nature, children are less prone to become overweight, and seniors live longer. But *how* this works is less clear. I'm torn between wanting to figure this out, and living in the moment as I walk this small beach looking at stones and birds.

Day 7 – *A quiet night, gentle waves, a few sprinkles, light winds, low overcast clouds.*

We wake early and break camp, grab a quick breakfast, load the canoe and are on the lake before the winds rise. We paddle north for three and a half miles to a prominent buttress, then move into high gear to cross the lake at one of its narrowest points. The mile-long crossing is uneventful, but the storms of the past few days keep us on edge until we are on the other side. Not a big crossing, just big enough to feel very alone in the center of the lake. The wind rises as we paddle south to camp at Moonlight Bay.

I like to ask people: When and where are you most alive? One of the answers I've heard repeated the most over the years is *outside, in nature*. That's true for me as well. What better way to fight apathy and despair than to get outside to discover the world around us?

Again, I come back to the question of how might we define and measure a dose of nature. Do people need large expanses of wilderness, or is a forest or small grove of trees enough? What about a garden? Is a mountain lake similar to a desert oasis? Each is a different experience that may provide different benefits. Nature offers many values: it provides shade in hot weather and solace in hard times. It offers beauty, breathing spaces, sacred places, playgrounds, and proving grounds. It inspires awe.

The sun is shining. Gentle waves lap the narrow pebble beach. The rocks sparkle, every pebble and stone shot through with gleaming bits of mica. This side of the lake gets full afternoon sun and several species of dragonflies hunt mosquitoes over the small cove. A quick swim in cold water makes me feel really *here*. What a difference blue sky and activity make in my mood. I feel healthy and fit. The lake is alive and so are we.

Varied, Hermit, and Swainson's Thrushes sing from the dark woods. The forest around our camp is a mix of life zones that, farther south in more familiar terrain, would represent different elevations. Here, Western Hemlock, Red Cedar, Engelmann Spruce, and Lodgepole Pine mix with Mountain Hemlock, Subalpine Fir, and Ponderosa Pine. Penstemon and saxifrage cling to rock cliffs above the water. Orchids, Twinflower, and Bunchberry cover the slopes above camp. As evening falls, whitecaps join the parade into the night.

Day 8 – *Clear sky at dawn, a light, transparent blue, almost without color. Small scattered cumulus clouds begin to rise around the edge of the lake. Still water.*

The call of loons wakes me in the morning and draws me out of the tent. Being on this lake makes me feel alert, confident, and strong. I understand the benefits of exercise, eating well, and a good night's sleep—all are part of this wilderness adventure. I'm curious about how often (frequency), how long (duration), and how many times (repetition) people need to engage with nature to get some of the benefits that I value so much. Some things are relatively simple to measure. Deeper impacts on body, mind, and spirit are more difficult to understand.

What are the characteristics of nature that impact human health? I'm most interested in the *quality and intensity* of the experience. What's the level of physical or mental exertion and challenge? Is it a solitary or shared experience? How does the place itself—the richness or complexity of the environment—influence body, mind, and spirit? Is it enough just to *be* outside or does it matter what you are *doing* while you're there? I believe the choices we make about what we do are critical: paddling a northern lake with a group of teenagers is not the same as watching moonrise by yourself from a mountain summit. Keying out a new plant or identifying a bird by its song is not the same as a seventeen-mile trail run. I'm not sure it's possible to measure or compare such disparate doses of nature, but nevertheless I think *how* we spend our time is one of the most critical aspects of our experience in the natural world.

Of course this could all be moot. We know that nature offers a rich variety of experiences that leads to healthy lives. But people must be able to choose how to take advantage of them based on their interest and abilities and for that to happen, nature must be available, accessible, and welcoming. That's not the case for many. We need more access to nature and for that we need more nature both close to home, and in the big wild far away—places like this remote, windblown lake.

Shelley walks the pebble-stone sparkling beach one last time before bed. I turn off the stove and pour a last cup of tea. Soft waves lap against the shore. My nine-day beard feels soft to my calloused hands. A trio of thrushes sings us to sleep.

Day 9 – *Clouds cover the lake at dawn and dissipate as sunlight touches the water. No wind, the lake is still.*

I drink tea and watch light descend the slopes of Central Mountain. Today's plan is to paddle ten miles south to Murtle Lagoon, an easy four hours, and begin our journey home.

Things do not go as planned. We leave Moonlight Bay in mid-morning, canoe packed tight and spray deck on, heading south into a gentle breeze. Almost immediately we find ourselves in three-foot breaking waves as whitecaps cover the lake. Waves splash over the bow of the boat and into Shelley's lap. After a couple of hours of this we call it quits, surf uncomfortably backwards into a small steep cove, and jump out to hold the canoe upright in the surf. We bail four inches of water from the boat. An hour passes and the lake begins to calm. We push off again. Within minutes the wind rises. We continue around one cove, then a long exposed stretch of rocky forest, our passage agonizingly slow.

The wind switches from the west, waves push us closer to the branches of fallen trees that extend into the lake like clutching hands. We pull into another small beach to stretch and bail. I hold the loaded canoe in place while Shelley wades and bushwhacks around the next point to see if there are better spots to camp or regroup. I watch loons dance in the moving waters and have just begun to worry when she reappears, wet to her waist, hair wild and eyes bright, to report that there is nothing ahead but more of the same.

Again the wind drops and we paddle on, past a long stretch of dense forest and rocky headlands. And again the wind rises and we retreat to shore, this time on a steep, narrow beach. Convinced we will spend an uncomfortable night here, we set up the tarp with just enough headroom to sit on the sand, and eat a bite of lunch. The wind and waves continue, and once again let up just enough for us to give it one more chance.

In early evening, exhausted and the wind still strong, we pull up to a long, flat beach at Straight Creek. I struggle with the flapping tarp while Shelley works on dinner. With the last of our red wine we raise our glasses to the lake. Five miles in seven hours. Damn nature! We hope for better conditions tomorrow.

All day I've been reminded how glad I am to be doing this with Shelley. If I was alone, in my solo canoe, I'd have spent much of the day terrified, and would probably be sleeping in an uncomfortable bivouac miles from here. Solitary experiences teach us how to listen and reflect. They provide additional levels of challenge but bring you a little closer to the edge. Shared experiences add multiple dimensions to the experience and help build community. One way is not better than another; we choose what we need each time we go outside. While the presence of others may distract or insulate us from the direct power of nature, shared experiences provide different opportunities for reflection, not to mention making them safer and more accessible.

Day 10 – *We wake at dawn to overcast skies and gentle waves.*

A couple in a red canoe heads south paddling fast. They yell across the water that they watched us through binoculars all day yesterday from their storm-bound camp on the other side of the lake. Like us, they are heading home.

We drop camp quickly, eat a couple granola bars, take a couple ibuprofen, throw on the spray deck, and are on the water by 7:00. The light breeze on our faces feels good—it tells us we are moving. The forest is still dark with shadows. Black rocks appear beneath the canoe. Endless shoreline. Wild country. A five-mile paddle takes us only an hour and a half on rippled water accompanied by loons.

I am trying to define something that we don't have a common language for—how we understand and share the value of nature. A dose of nature is a prescription, a quantitative measure of an essentially qualitative experience. What do we gain by measurement? What do we lose? Can we also measure dosage in terms of *stories* brought back from wild places that help us remember and make sense out of special times, places, and experiences? We might also ask how much *love* we need. For we need love just as we need the natural world. The real question is what do we need to be *fully alive?*

I believe that nature can heal us and heal our communities. Walking in the woods, paddling wild lakes, and practicing natural history are all ways we engage our senses and deepen our connection with nature and ourselves.

How many mornings do any of us have left? Why not wake early *every* morning and greet the day outside? We need to welcome sun and storms, loons and clouds into our lives.

Meeting the Gray Fox

PABLO DEUSTUA JOCHAMOWITZ

How does it feel?
How does it feel?
To be on your own
With no direction home
A complete unknown
Just like a rolling stone...

Bob Dylan

Paul arrives on time. I am struck by the slowness of his movements and the blankness of his face. I imagine this young man carrying an invisible backpack full of stones that curves his back. He takes the chair across from me. "I am dead inside," he says. "Nothing seems interesting, nothing captures my enthusiasm . . . I feel dead . . . I am dead and utterly lost." Hard words to hear, particularly from a young person, with all life and possibilities ahead.

Paul is twenty-five years old. He finished his university degree in political science two years ago with extraordinary qualifications. I ask how long he has been feeling like this, and his answer shocks me: "Since I was ten years old, I think." It is not easy to follow Paul's words: his somnolence and lack of energy spreads all over the therapy room and I also start feeling tired and sleepy. I make an effort, though, and keep enquiring about his life. What happened in your teens, I ask. "Nothing in particular," he answers. Everything was supposedly okay. He had friends, dated three girls, received good grades. "People loved me, everyone loved me. I usually heard I was a role model for everyone, that I always did what I had to do, the right thing, on time, the proper way . . . as everyone expected."

In our second meeting, I ask Paul about his last thought of the previous session—that he had always been a role model, and had always done what he was supposed to do. Something changes in the expression on his face. "I remember much of my childhood . . . I enjoyed a lot, I felt light . . . without pressure . . . free, flexible."

I ask, "What changed? You couldn't keep walking light . . . what happened?"

He keeps silent and seems anxious. I stay there, waiting for whatever he wants to share. "I started to feel extremely pressured to be someone . . . to be someone intelligent, to be the best. I didn't know what 'to be someone' meant, but I felt the heavy pressure on my shoulders . . . this pain in my back . . . this oppression . . . like a black hole inside my stomach. I didn't know what was happening to me. I was scared . . . I'm scared. I don't feel I am myself anymore. Who am I? Where is my essence, my essence...?"

His words are so intense and I feel something very real in his speech. It is like something deep within Paul awakens to life through those words. I hear a hint of a rebellion within them—a rebellion against what, I wonder.

I have no answers to offer, but I believe something needs to be said: "Paul . . . I have the feeling that you want to rebel against something or someone, you want to disobey . . . you need to break the rules and to be released from some kind of heavy chains." He remains silent. I can see him breathing uneasily, his gaze wandering rapidly from one spot of the ceiling to another. He nervously clutches the lapels of his shirt. "I can't take this anymore. I can't take this anymore." He is sobbing—an uncontrollable crying, wild, full of rage. There is pain in that cry, of course, but there is also truth and an honest connection with himself. I let him cry all the time he needs. I keep him company, silently. I feel words are not necessary.

Eventually, he continues. "Everything went wrong when I started growing up. I remember thinking to myself: 'I want to be a small boy forever.' But it didn't happen, of course . . . the child went away, sadly. I felt compelled to be the best, to be a winner. I couldn't fail. And everything I did . . . I do . . . it must be the best. I have no option . . . the best. Sometimes I wasn't able to start anything new since I was so scared of failing. It was very difficult to accept I was not the best. Everything was competition. My friends weren't my friends anymore . . . they were competitors and I was supposed to beat them. I was told I needed to get the best grades, that I needed to win. That was the only option. I was told the world is very competitive and I needed to be prepared. That was the only way I could get what I needed to get. And people will love you for what you get . . . success . . . success . . . the road to success. If you are not on it, no one will care about you. And I've been driving that road the best I can. And I'm supposedly driving fine. I am winning . . . but I can't keep with this. I feel I am dying. I am chained, gagged, numbed. This is not me . . . this is not my essence. And this is the only road I know . . . and I am lost."

When Paul leaves my office, I am bewildered. There is something—an inner voice of intuition—that tells me I will

need to approach him in a different way, not from my classi-
cal psychoanalytical technique: elaborating interpretations and
bringing theoretical explanations about his suffering will bring
no help at all. I am sure of that. I start feeling restless, so I
stand up and start moving my body . . . head, arms, legs. I feel
oppressed here. The air is not fresh; the space is so small, so
confined. I feel like a caged cat and have the urge to run, to go
outside. Suddenly, like a violent flash, full of energy, heat and
electricity, a quote from Henry Thoreau appears in my mind:

I went to the woods because I wished to live deliberately,
to front only the essential facts of life, and see if I could not
learn what it had to teach, and not, when I came to die, discover that I had not lived. I did not wish to live what was not
life, living is so dear; nor did I wish to practice resignation,
unless it was quite necessary. I wanted to live deep and suck out
all the marrow of life.

Then, I knew: we must leave the four walls of the office.
I will walk with Paul on a journey to find his "marrow of life,"
the essence that might enable him to reconnect with the person he truly is. We must go, like Thoreau, to the forest.

When we meet again I suggest a walk through the pine
forests of the Sierra de Guadarrama, an hour away from
Madrid, the city where we both live. Paul seems surprised,
clearly not expecting this. I think: Good start—surprise is
something he desperately needs! There is something fresh and
new in this proposal, something which does not fit the traditional therapeutic method. I keep thinking, rapidly: Leave the
traditional, he's been so traditional and he's suffered a lot. He
needs a rebellion, a good rebellion. I hear Lou Reed's words:

take a walk on the wild side . . . take a walk on the wild side. I really feel the forest will help Paul to take this step into unknown territory where he can find his own path.

I say to him, "We will just walk and look, we will move and breathe...don't *think* about it. You have thought too much for so long."

Three days later we meet in Cercedilla, a small village at the foot of the Sierra. I deliberately didn't state instructions for this "session," so we silently start walking towards unmapped territory, inner and outer. We must not get anywhere—no goals, no time, no destination. It's only us and the rumors of nature. I can feel Paul calm, watchful and curious. His face shows a childlike expression, a mixture of naughtiness, excitement and fear. I tell him to let all control go away and to feel free to do whatever he wants—in the way and time he wants. He nods. The forest greets us.

We walk uphill, under the towering Scots Pines. A gentle wind blows, opposite to our faces. It's October and this cool air is welcome after the warmest summer in thirty years. The absence of human-made sounds (words, whisperings, music, cars, different machines) is extraordinary. The rhythmic sound of our footsteps on the carpet of pine needles becomes absolutely hypnotic. Paul stops, always silent. Slowly he drops to his knees and grabs a handful of needles with both hands. Eyes closed, hands to his face. I can hear him breathe deeply: one, two, three, four times. And again, this time even more deeply: one, two, three, four. He opens his eyes, and says, "What a smell. It moves me inside . . . so fresh and clean. My lungs are expanded as much as they can be. There's something good here, within this scent. God . . . I'm breathing . . . I'm breathing."

Little has to be said from my side. I don't want to break

his moment with words and thoughts and audacious psychological interpretations full of intellectual knowledge. No, I'm positive: this is another universe. Under these pines words just don't work as they do in the therapy room.

He keeps silent, not kneeling anymore but in a more comfortable position, half lying. I decide to sit, a couple of meters away. The sounds of the forest wrap and haunt us. The wind blows through the branches of the pine trees, as if they were rustic guitar strings, creating a very special melody. The Short-toed Treecreepers blithely sing. I watch Paul staring at a column of ants climbing the trunk of a pine tree. I think to myself, no one judges you here, Paul, no one expects anything from you. Feel free and light. You don't have to prove or do anything special. You just need to *be*. So simply *be*, my friend, as the Short-toed Treecreeper is *being* now, so distant from ideals and ideas about what *should be* done. It doesn't matter: the treecreeper just lives and takes everything as it is. No judgment. No assessment. No rationality.

I am about to explain this idea to Paul but, happily, I refrain. Stop, I say to myself. We are not in the therapy room. This is wordless territory.

Paul is looking, smelling, touching, listening, being . . . just like the Short-toed Treecreeper. That's all he needs now. Let nature heal him. You don't need to say anything, I tell myself. You don't need to play the brilliant psychotherapist. *Just be* here and witness this healing encounter between Paul and the forest.

Paul's words start to flow: "I am feeling different . . . it is an odd kind of feeling. I am a bit scared, anxious because this is so unknown. But I am okay with this . . . this is fresh air. I need to get used to breathing this fresh air. I'm not used to it and . . . well, you know . . . this fresh feeling of nobody expecting anything from me. I always have to perform in order to satisfy everyone—not anymore. I'm not feeling that here, in these woods. It is so easy. I don't need to act, I just need to let this happen. Look at those ants going up and down . . . I had this idea: they will keep living and doing their ant stuff with no care of what I perform or don't perform. I know it is very difficult to understand, but it feels so liberating, so true and honest."

Once again, nothing to say from my side. I just need to allow the landscape to keep the dialogue with Paul, with its own order and flow.

After a few minutes, he speaks softly. "I can hear a murmur in the distance . . . I've just realized that." He shows a deep concentration, as if the murmur was the most important thing in the world to pay attention to. "I believe it is water," he concludes. We rise and walk following the water's whisper.

Now, we face a creek that descends from a hillside full of pines. The crystal water hits against the rocks, forming eddies and foam. I think to myself, this is meltwater, coming from five hundred meters above us. I can feel the chill and the power this water has to awaken anyone from the deepest sleep or drowsiness. I mentally visualize Paul—the Paul who came to see me at my office the first time—diving within this chilly element and calmly waking up from lethargy, coming back to life.

The creek roar surrounds our bodies. We remain silent, mesmerized by the singing and dancing of water among the stones. I had never felt this intensity within the walls of the therapy room and I know nothing could be better now than this wordless moment among nature.

Time has stopped, one more time. Paul speaks, but I feel he is not addressing me, but the whole landscape. "I will never see it again—that small leaf the water is dragging . . . never. It is gone now."

I doubt if I should say something. I am not sure if I am really following what he wants to express. But suddenly, an idea shows up and I feel I need to give it voice: "It is true, more than true. Never again. Never again. That leaf is gone, forever . . . forever. Can you see it? Do you realize how special, how unique that moment was? What a gift! You were the only witness. You and the trees around us. And you are not asked to give anything back, you are not required to pay for it or to perform any role. You don't have to pretend to be something you're not. You just have to be here, and look, and smell and listen. Just that."

The words are coming from my mouth with no thought, with no aim. I trust they must be said. Like Paul, I am perplexed by my feelings. I have been trained to think, to understand, to analyze as a therapist in order to help my patients. But this is another story. This has nothing to do with analysis, thinking, and the rational mind. These are the Scot Pines, and the Short-toed Treecreepers, and the group of ants, and the creek acting as medicine, helping Paul and helping me. Humbled, I have utterly abandoned the pressure to act as the almighty therapist who must guide his patient to health. I feel relieved, light and free, as Paul feels. We are two simple human beings letting the forest cure us.

The following week, Paul opens the door to my office five minutes before schedule. I feel he is restless, maybe anxious. But I sense a different kind of anxiety this time. There is something new going on with him—I can feel vitality and eagerness to start talking. "Something has happened since we were in the woods last week. Something new—so new it makes me feel like I have to learn everything from the very beginning. Actually, it feels like I am starting to know myself from the

very beginning, from my true beginning, from my essence."
While he is speaking I recall the water roar in the woods, and
I feel he is roaring with life too. "I feel . . . I feel I will start to
see myself direct in the eye. No more stories, no more lies to
myself, no more." He keeps going: "This backpack full of heavy
rocks I have been carrying for so many years . . . well, it is time
to take the rocks out of it one by one. I still have plenty of
them, but I have already taken a few out. So difficult to express
with words, this feeling . . . this feeling . . . I don't need words."

Never before in my experience as a psychotherapist had
I witnessed such a radical change. And not just his words—I
was deeply impressed by the look on his face, the rhythm of
his breathing, the whole of his body language. Every feature
communicated lightness, open space, fresh air, freedom. I tell
myself: that's my job—to help my patients find their own way
to feel free and to live with lightness.

But in Paul's case I am positive it was the forest—not
me—who helped him.

What happened to allow this spectacular release of his
emotional burdens? One thing is obvious: direct and attentive
observation of nature was key. My mind wanders: Nature . . .
wild nature with its own laws, with its own order, so different from the regular way we use to lead our lives—speeding,
planning, being productive, self-controlled, rational, analytic,
competitive . . . such a different order! Nature has a different
way, a wild way. Wild nature. The word *wild* rumbles inside my
head and heart. Then, I remember Gary Snyder's *The Practice of
the Wild*. I rush to my library and take the book from the shelf.
Letters and phrases dance in front of my astonished eyes.

Everything fits now. Like a solved puzzle, Snyder's words help me come to understand what went on with Paul last week in the woods: "Nature is not a place to visit, it is *home*." For both Paul and I, our walk under the pines was a return home, to our true instincts, our true nature. One thing is doubtlessly true: our origin as humans is in nature, not in culture, and this is a forgotten fact—or repressed, as psychoanalysis would state.

I clutch the book, full of emotions. Among Snyder's descriptors of "wild individuals" are: *free, spontaneous, unconditioned, expressive*, and *playful*. I feel so connected to Paul while reading Snyder. I am picturing Paul's suffering all these years: he was deeply wounded by the imprisonment of his wild being, which for so long he felt chained, domesticated, forced to follow a set of rules and precepts that were not his. The *shoulds* and *musts*, the pressure to be a role model, suffocated him. Our walk under the pines—our coming back home—unlocked the door and released Wild Paul.

The adventure continues for both of us. We will keep going home, to the woods, in search of our "marrow of life." We will keep visiting the Scot Pines and the Short-toed Treecreepers to cultivate our wild essence, which, according to Snyder, is like "a gray fox trotting off through the forest, ducking behind bushes, going in and out of sight." Paul's gray fox, and my own gray fox, appeared suddenly and charmingly that wonderful day in the forest and became our guides.

Let's each go to the forest—silently and in awe—to find our own gray fox. Our true essence and health is there, waiting for us.

Environmental Generational Amnesia

PETER H. KAHN, JR.

Prelude

As an adolescent, I often rode on horseback as long and hard as my horse and I were able. I was living then on this land, 670 acres, twenty miles up a dirt road from a small cowboy town in the mountains of northern California. It was a commune back then, in the 1970s. Now a few people live here full time; the rest of us come as our schedules allow. It has become three generations of us.

On one of those rides, I was with my friend Hawk. After a week, we were exploring some country on the far side of a wilderness area. We camped in a meadow below one of the big peaks. There was lush grass for the horses, which they appreciated. We let his horse roam freely, as she stayed reasonably near us for the night. But my horse, Val, tended to wander longer distances, and so I hobbled her. You can think of hobbles as like large handcuffs—"hoof cuffs"—that you put low down on a horse's front legs, with the rings of the cuffs being comprised of two-inch flat leather straps. To move, Val could take tiny baby steps. Or she could jump off with her hind legs while throwing her front legs in unison forward, which is how she usually did it. That jumping was inefficient and hard work. The idea was she eats the grass all around her, takes a jump, and then eats more, and so on, and by morning she has eaten well and is not too far from us.

But the next morning Val wasn't anywhere to be found. Hawk and I walked a large circle around where we camped, and then a larger one, but didn't find her. We then hiked both directions on the nearby trail for about a half mile, but no signs there, either. We figured we had a bit of a problem because, I mean, I had lost my horse. We were about forty miles from home. We looked at each other, and said, well, let's track her! Which was a good idea except neither of us had learned how to track. Still, we figured the basic idea was like Hansel and Gretel. Find the first one, then find the next one, and keep it going. It's much harder than it sounds. Our success had to do with us discovering, at two critical spots, not her tracks, which we hardly ever found, but horse poop. If it wasn't for that, we would not have continued venturing on the line we were on, because we were heading further and further down into a steep rugged ravine, and it was clear to us that there was no way out of it. But we kept heading down, and there we found Val on a spot where it was too steep for her to go further. The distance itself was impressive. We must have been a mile from our camp, and she had hobbled it all. But what puzzled us was the direction. Why this way? It was so steep, and it led nowhere.

When we got back to our saddles, we pulled out a map. And then the answer seemed clear. For the entire week we had been making a large loop from the ranch into and around the wilderness area. We were about three-quarters of the way done with the looping part. Now, when we drew a straight line on the map from where we were back to the ranch, that's the exact path that Val had been on. She, like us, had never been in this area before. But it seems that she knew the most direct way home, and was trying to get there as if she were a homing pigeon, not a four-legged animal.

Sometime later, I sold one of my horses to Hawk, who was seventeen years old by then. I sold that horse to Hawk for a small price, as she seemed to me a little round, out of shape, and never inclined to get in shape. But Hawk found in her more than I had ever seen. One day he saddled her up, and rode north from our ranch. He rode for about eight hundred miles. He had wanted to ride to the northern part of Washington, and he did.

Mind and Body

I and others came of age here with our minds and bodies living on big land. It's deep in my psyche. And something like this is deep in the psyche of all of us. Because for tens and even hundreds of thousands of years, we as a species came of age on the savannahs of east Africa. That land had even more wildness to it, not defined as pristine or untouched by humans, but in terms of it being large, unbounded, mostly unmanaged, self-organizing, and unencumbered and unmediated by technological artifice. We're drawn to that form of wildness. We love it, and we can fear it, too. We need deep and intimate connection with nature, and more wild nature, to do well physically and psychologically. The argument and evidence for this proposition is sometimes brought forward under the umbrella of what E. O. Wilson and others refer to as *biophilia*: the innate disposition to affiliate with life.

But if it's true that we need deep and intimate connection with nature, then a puzzling question emerges: Why are we degrading and destroying nature at such an astonishingly quick pace?

I first began to recognize the problem when I was interviewing black children in the inner city of Houston, Texas, about their environmental views and values. In some respects, these children articulated surprisingly rich accounts of their interactions with, and indeed moral regard for, nature close at hand. But I was especially surprised by one finding. A significant number of the children interviewed understood the idea of air pollution; but they did not believe that Houston had such a problem even though Houston was then (and remains) one of the most polluted cities in the United States. I would wake up in the mornings there stifled by the smells from the oil refineries, and my eyes would sting a little. How could these kids not know it?

One answer is that they were born in Houston, and most had never left it; through living in Houston they constructed their baseline for what they thought was a normal environment, which included that existing level of pollution. When they did speak of air pollution in their own city, they would mention, for example, occasional fumes they could see and smell behind some diesel buses.

Building on this research from the 1990s, I proposed that people across generations experience psychologically something quite similar to the children in Houston: all of us construct a conception of what is environmentally normal based on the natural world we encounter in our childhood. The crux is that with each ensuing generation, the amount of environmental degradation increases, but each generation tends to take that degraded condition as the non-degraded condition, as the normal experience. This is what I have been calling *environmental generational amnesia*. It helps explain how cities continue to lose nature, and why people don't really see it happening, and to the extent they do, they don't think it's too much of a problem.

Some years ago, I met with two architects from an international architectural firm. We were considering a collaboration on some of their projects, where I would contribute my perspective of how to design spaces to increase people's direct and sensorial interaction with nature. I now call this approach Interaction Pattern Design. I see this approach helping to counteract people's impoverished experience with nature, and thus trying to shift our environmental baseline the other direction.

The architects understood about environmental generational amnesia conceptually, and they could see how it was taking hold in the younger generation. One of them, for example, said that while he was driving with his son down a street in Seattle, his son said something like, "Dad, look at the beautiful forests!" And the architect laughed with just a little sadness, because he knew that his son had come of age in the city and had constructed a concept for what counts as a healthy forest: healthy trees lining a city street. The architect asked me, "That's an example of environmental generational amnesia, isn't it?" I said, "Yes, that's a good observation." But what he said next was typical, and endemic to the problem. He said that thankfully he, himself, didn't have too much of a shifted baseline. He knew, for example, that forests weren't the tree-lined streets, but rather the large forested lands when you drive east out of Seattle toward Snoqualmie Pass. I didn't know how to tell him that almost all of that land has been logged many times in the last 150 years. We've lost the perception of old-growth forests, and think of logged land as the pristine, and the normal. In their journals, Lewis and Clark wrote of seeing in the Missouri

River bottoms no less than ten thousand buffalo within a circle of two miles around. That's a lot of life! Now we might see a few deer in the logged forests and think "that's a lot of life!"

I've conducted research in related areas to try to make the ideas more compelling to a Western mind that seeks scientific evidence. In one study, for example, my colleagues and I engaged ninety participants in one of three conditions in an office setting. One involved a glass window that overlooked a beautiful large fountain and an expanse of grass, trees, and sky on the University of Washington campus. In the second condition, we inserted a large digital display into the window, and then mounted a HDTV camera above the outside of the window, so that the window displayed the *real-time* digital view of the same nature scene. In the third condition, we covered everything up, and made the office functionally into an inside office. Over repeated times across each condition, we induced low-level stress to the participants, and then measured heart-rate recovery. We found that, in terms of heart-rate recovery from low-level stress, the glass window with a nature view was more restorative than an inside office. Then we asked whether we could get similar heart-rate recovery by employing *technological nature*: the real-time digital display of the same scene. The results showed no. Indeed, participants' heart-rate recovery did not differentiate statistically between the technological nature window condition and the inside office condition. In another study, we inserted these technological nature windows into the windowless offices of seven faculty and staff, and then over a sixteen-week period, we assessed participants' practices, judgments, beliefs, and moods. Here we found that participants enjoyed the technological nature window, and benefited from it in terms of their psychological well-being, cognitive functioning, and connection to the natural world. These two studies taken together convey the basic finding from my other studies that employ other forms of technological nature (such as robot pets).

The overall finding is that interacting with technological nature is usually better than nothing, but not as good as actual nature. How does this relate to environmental generational amnesia? If we evaluate the effectiveness of technological nature compared to no nature, we will falsely believe that we can destroy actual nature with psychological impunity. We'll "forget" experientially what the experience of actual nature feels like, does for us, and with us, and normalize to the new diminished technological baseline.

The land above my cabin had been old growth when I was an adolescent. Then Georgia Pacific logged it. I had no idea they could do that. I was learning about the world. Some years later, they logged it again. Then they sold it. The person who bought it then logged it. Over the span of my forty-five years here, I've lost track of how many times that parcel has been logged. I think six times. Spotted Owls used to hoot—*whoo, whoo, whooooooo*—and fill the evening air with their presence. But after the land was logged for the third time, they all left or died. Spotted Owls are referred to as an umbrella species: when they go, so do many other species. None of the younger generation on this land today misses the Spotted Owls because they don't even know they existed. Similarly, none of the younger generation here even considers saddling a horse and riding north eight hundred miles to northern Washington, as Hawk did—in part because it can't be done anymore. Too much nature has been populated, and land further subdivided and fenced. Thus our environmental imaginations have also been diminished.

Woodland Park Zoo in Seattle keeps captive an enormous Grizzly Bear in a space the size of a McDonald's parking lot. The grizzly has been forced into a lifetime of solitary urban confinement. Some of the students in classes I teach at the University of Washington have told me that the Grizzly Bear in the zoo teaches us to love wild animals. But if we loved wild animals, how could we not cry in the face of such captivity? I think it's because we have caged ourselves within the boundaries of our cities.

Sometimes colleagues tell me that there isn't a problem. They say that we evolved as a highly adaptive species. And that we're adapting to more urban life, less nature, and more technology. It's just different now, they tell me, neither better nor worse: "We're changing, enjoy it!" And, yes, there are things to value greatly about cities, especially as they bring people together, facilitate the intermixing of ideas, and nurture a huge creative flow. But urban density is coming at enormous costs to our physical and psychological well-being.

We're adapting but not flourishing. In the United States, for example, about two-thirds of the population is overweight, and one-third is obese. How can we look at those numbers and think, "oh, that's just normal." No, it's not normal. How can we accept it so easily? It's not a healthy way for people to live, in body or mind. In the United States, ten percent of people take antidepressants. It's the most prescribed drug. Eight percent of people in the United States have asthma. The numbers must be much higher in polluted megacities like Beijing, Shanghai, Sao Paulo, Mumbai, Kolkata, and Mexico City. The number of such megacities is expected to rise substantially over the decades ahead. Let me say this a different way. A lot of people can't breathe normally. This is not just us adapting and doing fine. Elephants in zoos can be seen stamping their feet for hours on end. They have "adapted" in the sense that they are living, but they are going crazy—in a literal, psychological sense. I remember being at Brookfield Zoo in Chicago, and there was a large African elephant there that the zoo named Happy. I was standing by her exhibit (should there even be language of having animals on "exhibit"?), and I heard a woman say, "Oh look at Happy, she looks so happy to me!" We no longer know what we're looking at. In the United States, nine percent of people have diabetes, twenty-five percent of people experience a diagnosable mental disorder in a given year, and fifty percent of people have one or more chronic health conditions.

Every time I return to this land, it's like I am able to breathe again. I lie on the ground at night, looking up at the stars, such vastness unending. Our bodies and minds were made to look up at the night sky. They were made to walk long distances. They were made to swim in wild rivers, to pick berries, and to look into the eyes of a wild animal for a moment, and to recognize the consciousness of what Paul Shepard calls the Other, and to know that that animal is recognizing your consciousness, through their own form of it. That's part of the lovely reciprocity with nature. It's as similar, different, and beautiful as when you look into the eyes of your beloved, and it's reciprocated. To see and to be seen. To know and to be known.

Spirit

Over the last twenty-five years, I've been writing and conducting research on the importance of nature in human lives. But I had been leaving out of my professional work what I think is the deepest and most important reason of all, as it has little to do with reason at all, and is thus hard to explain. More recently, though, I've tried with such diverse groups as cognitive neuroscience roboticists and behavioral neuroscientists. To them, I have said that in the *Tao Te Ching*, written over 2,500 years ago, the Chinese philosopher Lao Tzu wrote that a cup's existence depends on both what it is (its objectness) and its emptiness (its space). A cup without empty space isn't a cup. Thus things are ontologically constituted by what they are and what they are not. To them, I have said that consciousness is not an epiphenomenal illusion caused by chemical processes in the brain; it's the state of being aware of not only the manifested form of the world (including one's thoughts) but the formless.

Tao refers to the Way, the source of all existence, the all-pervasive substratum that has no physical manifestation. Tao is a transcendental reality. Verse 4 of the *Tao Te Ching* (Trans. J. Star) begins:

Tao is empty
yet it fills every vessel with endless supply
Tao is hidden
Yet it shines in every corner of the universe

How did Lao Tzu know this? How do we know this today? That's the epistemological question for the ontological claim. The answer is that such knowledge comes to us not by modeling the brain or FMRIs, or scientific experiments. Indeed, it's not even through words. In Verse 1 of the *Tao Te Ching*, Lao Tzu says that the Tao that can be spoken of is not the true Tao. Words are at best pointers or sign posts. We begin to sense this truth through the space between words, between ideas. Through stillness of mind. Some people realize this space through meditative practices, and that's good. But nature, and especially more wild nature, is itself a portal to this transcendental reality. When, for example, during that moment you look into the eyes of a bear, and the bear is looking in yours, thought disappears. Yet you are highly conscious. More conscious, in fact, than when your mind is ruminating endlessly about things that happened yesterday or in your childhood, or about future problems you're trying to solve that have not even become problems yet, except in your mind. When you put your hands on your dog, there is often a little of that space that emerges. The effectiveness of human-animal therapy builds on this formless dimension. You can also put your hands on an old-growth cedar that rises high. Perhaps also allow your forehead to be in contact with the bark. In that position, be still for ten seconds. Not still just with your body, but also with your mind. Be aware of the space around your thoughts. Through that space and with your body, at least I have found it so, it is possible to access a little of the consciousness of the tree. Maybe you've lived with someone for a year or twenty years, and you think you know them. But there is more. We see only the surface and just a little below. And when some people see a little deeper than that, they say that everything is connected, and that there are no isolated or meaningless events.

I wonder if it's possible that in the continuing evolution of our species we are coming to the end of a highly dysfunctional state, which is discontinuous with what will follow. The analogy I've heard is that we're like a caterpillar barely able to crawl further as it prepares to become the butterfly. I would like that to be true. If it is, I would like there to be some nature around when it happens.

Sensing Place Before and After Race

LAURET SAVOY

Sometimes I awaken at night, disoriented from the same dream. *I breathe but I'm not alive. I breathe but I'm not yet alive.* Then the dream crosses a threshold between two memories: a five-year-old daydreaming in slant golden light and an eight-year-old praying for yellow. Three intervening years will alter how this young girl measured her known world.

"Expose a child to a particular environment at his susceptible time," Wallace Stegner wrote in *Wolf Willow*, "and he will perceive in the shapes of that environment until he dies. The perceptual habits that are like imprintings or like conditioned responses carry their habitual and remembered emotions."

My life took distinctive shape within the frame of coastal California in the 1960s. I was born there, the only child of older parents who had come west in search of a new chance. This would be my father's last attempt to build a life far from his familial home of Washington, D.C. Because of him, there was frequent movement and change: from San Francisco to Los Angeles, from bungalow to apartment to second-story flat on Redondo Boulevard. And, because of him, I learned to perceive by the textures of dry lands at my susceptible time.

The dailiness of our lives took place within the City of

Angels. Yet what lay outside urban edges mattered far more to me. Southern California's sky and intensely physical landscape became reliable companions in a neighborhood with few children. The San Gabriel and Santa Monica mountains formed the rugged northern horizon. Santa Monica beaches and Pacific Ocean Park lay to the west. Any excursion to the coast, mountains, or beyond imprinted in memory, like the day we crossed the San Gabriels to the edge of the Mojave Desert, to picnic at the Devil's Punchbowl.

But foremost was the quality of light, its depth and brilliance, such that by the time I turned five I had defined who and what I was: a child of my parents, yes, yet also born of sun and sky.

Facing west from my bedroom porch I liked to watch late-day light paint the sky, paint nearby homes, and paint me. One memory still remains sharp, where I was, the time of day. Golden light embraces everything, my skin color taking on the cast of sunlight. I believed then that, yes, sun made my skin. And at the same time I am looking north, seeing the hills, seeing them take on that same cast of light. It's as if we are twins, the sun's warmth coloring the land, too. The northern sky is deep blue and I look again at the inside of my arms and see my veins coursing with blue. There could be no other way that I came to be—beyond being a child of sky, a child of sun. If I heard the word "colored" then, this would be its meaning.

But my father decided to return to Washington, D.C., in search of employment and dignity. Because a seven-year-old has little choice in such matters, short of running away, the only option was to bring home with me on the drive east. I tried—but how does one hold sunlight or keep ocean water from spilling or evaporating?

As the distance from California grew, so did my vision

of home. Postcards of Pacific beaches, coastal mountains, and the Mojave Desert were soon joined by images of the Sierra Nevada, of Zion, Bryce, and the Grand Canyon, of the Kaibab plateau forest, of the Colorado River and Rocky Mountains. Stones from these places came east, too.

🌿 🌿 🌿

Though riots had erupted earlier in Watts, though the civil rights movement was in full stride, I knew nothing of "race" and human discord before the move. Light and textured geography defined the California I knew. Race would find me in Washington, D.C., after the 1968 riots.

So, a continent away from home, a girl who believed she came from sun and sky now prayed that all daffodils were yellow. March had begun to brighten winter's fade in the capital city's corners and shadows, trees and grass. But along the school playground fence I couldn't be sure what color our garden would grow. Months before, in October, we knelt aligned with chain-link—ten, eleven, twelve second-graders in dirt. Digging, careless of school clothes. Digging, laughing, squealing—two bulbs each. Digging to Sister Mary Richard Ann's *gently now, children, gently.*

Until one boy sang out, pointing at me, *nigger flower nigger flower, ugly dark and dirty flower* . . . ignoring sister's hush. I filled five months with pleases. Please to the unborn plot, please to poking greens, please to suspect papery wrappings. Please. This, too, was a child's susceptible time.

Writing from the vantage point of retrospect across long years has its hazards: a full-bodied mosaic of experiences becomes inscribed in the mind as fragmented, condensed horizons called memory. I recall first recognizing, as a seven-year-old, racism aimed at me in spit and curses in a city recently in flames. I recall that it was easy to imagine even humidity-faded sky born of unfathomable hatred and decay.

Wondering if I should hate in return, I learned a crucial lesson. The land did not hate. People did. Refuge seemed to lie in wild places, in national parks of the West. But my one desperate wish couldn't be granted. The constant question to Father, "When are we going home?" always met the same response: "We are home."

Yet a turning point hinged on this eastern landscape. Even though the West I yearned for lay impossibly out of reach, I could still escape erosive moments here, outdoors in nature that didn't judge me or spit. The Potomac River. Rock Creek. The Blue Ridge in Shenandoah National Park. Civil War battlefields: Manassas, Antietam, Gettysburg, more. In these places, I began to realize that Earth was much older than humans, that the antiquity of the land itself preceded all human ugliness, even the violence of war and riots. By the age of nine, I began to reach toward Earth's deep time for solace.

What I couldn't fathom then was that my turn to this humid land's past began a turn toward blood origins. My parents had told few stories of family ties to any place as home. Shaping contours of generations were lost in silences. Only in recent years have I come to realize that familial roots grew in the Chesapeake's tidewater rivers, lowland coastal plain, and rocky Piedmont.

Tribal peoples—Piscataway, Powhatan, Nanticoke, Rappahannock, Mattaponi, and so many more—long claimed this area as homeland. Yet by 1617 English colonists began to found an empire of "smoke" after discovering the marketability of tobacco. Enslaved Africans soon powered the cultivation of Virginia and Maryland's export staple.

As peoples from Africa, Europe, and Native America converged in the seventeenth century and later, their interactions with and experiences of each other took many forms. Conflict, dispossession of homeland, slavery, and near-bondage occurred with collaboration and inter-"marriage." Later, descendants of colonizers, the colonized, and the enslaved would also come together and build the nation's new capital, carved from Maryland and Virginia along the Potomac River.

Bloodlines leading to my father, and to me, originated on three continents and joined here over centuries. These ancestors' lives were framed by the lay of Chesapeake land and rivers, by soils and their exhaustion, by colonial perceptions of nature harnessed for profit, and by the establishment of the nation's capital in a region condoning chattel servitude.

California was my birthplace, the place before race where impressions of land and sky took deep formative hold. Our move east at a time of riots seeded my need for grounding beyond race in the absence of family story or direction. Western earth and sky, my first home, will always pull—I still prefer the textures of dryness to humidity. My gaze still turns toward slant, late-day light. And I eagerly await the spring bloom of daffodils. But, now, as I uncover more ties to the Chesapeake tidewater and Piedmont—sites of familial memory—I'm finally coming home to a larger sense of how ancestral lives *took place* in this land.

Dark Love

JANA RICHMAN

In a dark time, the eye begins to see.

Theodore Roethke

It slips in quietly. A hint of terseness marks his voice, an opaque film covers his blue eyes, his face flushes and its lines deepen. His 6'4" frame droops toward the floor as if he's ashamed to drape his sorry self over it, and he tries to creep from the room unnoticed. It hurts him to be seen.

We share the only bed in our house, but he curls close to the edge, his face in the moonlight twisted and consternated. I want to reach out with a soothing touch, but I have learned not to. When he is deep in his dark world, a simple touch will send a startle response through his bones. He will burst from the bed as if facing a knife-wielding attacker and his wild eyes will be locked on me.

When I wake in the morning to find his side of the bed cold, I search for signs: a spoon in the sink indicates coffee was made; a creaking floor in his upstairs office indicates movement. From the signs, I can measure the depth of his depression and the probable length of its stay. No signs at all, and I feel as if I've been stalked into a dead-end alley.

I once believed myself capable of empathetic greatness, a belief that's been gutted and redesigned like a nineteenth-century farm-house. The crumbling bricks still hold, but the interior structure bears little resemblance to the original.

☙ ☙ ☙

Steve was fifty when we met; I was forty-eight. Our future held no golden wedding anniversary; silver was dubious. Such reckonings cut short the discovery period of romance enjoyed by the young. We acknowledged our love for each other, and, almost in the same breath, we acknowledged our impediments: Steve's depression, my anxiety.

Having anxiety in our anxious culture is like wearing a white t-shirt—it's not conspicuous—so I had minimal aware-ness of its scope. And being wholly naïve about depression, I shrugged it off in the name of love. With less caution than warranted, Steve and I joined hands and stepped into the abyss.

Anxiety and depression share commonalities. In our case, the emotional memories of each are decades—maybe genera-tions—old, with no faces, no bodies, no specific points of ori-gin. The similarities generate compassion between us but not necessarily understanding. And distinct differences make us ill-suited for sharing a life.

Anxiety gushes out, soliciting reassurance and relief; depression pulls in and sets up barriers. Anxious people want to process, often in a desperate, frenetic way. But insisting that a depressed person process his current state is worse than futile; it is merciless. Working together, depression and anxi-ety construct a near impermeable trap. When I sense Steve's depression, I churn in angst. When Steve senses my anxiety, he drops deeper.

Steve's depression is episodic, triggered in a moment that takes him down. And in that moment, life is brusquely shifted, shut down for an indefinable period. When I first saw it, although I had been forewarned, I had no idea what I was seeing. The shift in his physical appearance alone pulled me up short, and the abrupt change in personality seemed like a subterfuge. And for many years I treated it as such, demanding that he stop and explain himself.

❦ ❦ ❦

He retreats into his impenetrable misery behind the closed door of his office. I walk to keep my body occupied while my emotions lurch from confusion to sadness to anger to desperation. I return to a quiet house, no traces of movement. I search the bookshelves and Internet for comfort. So much advice—all of it familiar, none of it useful.

Two days go by without verification of life. I stew and listen and watch. I dissect the days and hours leading up to the moment it slithered in. I pinpoint the trigger and rewrite the script. I chant a whispered mantra: This will end. *But I worry that it won't end, that we'll be here on our respective sides of a cheap, hollow door three weeks, three months, three years from now.*

On the third day, the door opens and I jump to attention. He slouches down the stairs without making eye contact, looking ten years older than he looked four days prior. I offer to make soup, I suggest a hike, I extend bookshelf advice in a cheerful voice tinged with urgency. I speak to him as if he doesn't understand his own mind. He goes back upstairs and shuts the door.

❦ ❦ ❦

Steve embodies light and dark in their extremities. The dark runs deep and murky, but radical light runs parallel. I fear the dark will snuff out the light and destroy him, destroy us. He assures me that will never happen, and like a religious skeptic

209

teetering on the edges, I work to keep the faith.

I want to pry him apart, separate light from dark. I want the model with the personalized options, not the package deal, but his GPS is already installed. Ripping it out would leave him lighter, yes, but also deformed, shrunken, misshapen. Much of his beauty comes out of the shadow. His gentleness, his patience, his wisdom, his passion—all flow from having dwelt in the tender place of despair. I deeply understand the truth of this. Still, I want it to be easier—for him, yes, but mostly for me. He knows this darkness, and he oddly draws strength from its familiarity, as if it constitutes some sort of sacred ritual. I cower in its presence.

❦   ❦   ❦

On the fourth day, I wake to find the office door open and him gone. I breathe a sigh of relief for a morning without his dark presence and say a small prayer to the gods he worships: redrock canyons and sagebrush flats. He has gone to the desert.

I walk out to the garage to see what's not there: a cot, a sleeping bag, a five-gallon water jug. All good signs. He will spend nights under a dark sky, and when the sun rouses him, he will walk between redrock walls, bumping against them in his rawness. He will find a flat run of slickrock to lie upon, and he will stay until desert light finds a fissure in his constructed shield. Then he'll come back to me.

❦   ❦   ❦

Shortly after I met him, Steve said something that would become a refrain in our relationship: *I need to go to the desert.* We met in Tucson and lived in Salt Lake City, so technically we had always been in a desert, but that's not what he meant. He sought a desert free of humans and their debris, full of light, where he could dwell undisturbed for an extended period of time.

Having grown up in Utah's West Desert, I, too, have an appreciation for such places, but I initially thought him prone to hyperbole. Imprudently clinging to the popular view that all power lies within, I equated Steve's stated need to the exaggerated notions of a teenager needing a new iPhone. But after twelve years of inadvertent research, my flippancy has waned.

On our wedding day, Steve promised to always rescue himself—it was written into the vows. In my most anxious moments, I have extracted the promise from him again and again, but the last time I did was in the autumn of 2013, which was when I, at long last, understood that he has only one failsafe rescue: the desert.

It was our worst year together, high anxiety and deep depression, each tightening the knots of the other. We futilely tugged from opposite ends for eight months. In the fall, I suggested a weekend backpack on the Escalante River, and he nodded his agreement. But on the day we were supposed to leave, he couldn't rally the energy to abide my company, having, no doubt, sensed my desperate reach for relief. After he shut the upstairs door, I sat amid the mess of freeze-dried food packets and cried. Then I packed.

I would like to say I left the house quietly, but I didn't. I breached the sanctity of the closed door and made a dramatic, sobbing speech and exit. I no longer remember the words, but I remember the cruelty behind them. I'm sure I demanded some sort of promise or explanation that he could not possibly give. I remember his horrified face as I loaded my pain onto his.

I drove fifteen miles to the trailhead shaking with the kind of generalized rage that has no receptacle. Only after hoisting the pack and splashing through the knee-high, sun-warmed water for the first of many river crossings did I acknowledge that I had never backpacked alone, never spent a night *out there* by myself. It was an easy three-mile hike upriver to the Sand Creek confluence where I planned to camp, and the physical risk was minimal. But the sun drops early in the river gorge, and the long stretch of night ahead played on my nerves.

Righteous indignation propelled me forward, a feeling of something having been thrust upon me that I did not deserve. I slogged through deep sand, stumbled often, and expended a great deal of energy to gain little ground. Had I lifted my eyes from the trail, I might have been awed by Escalante Natural Bridge, a sturdy, flat-topped, deep red and brown arch that spans a side canyon like a train trestle. Had I lifted my eyes, my heart may have been lightened—or at least distracted—by the Indian domicile ruins on a ledge next to a wall of seven-hundred-year-old petroglyphs. But I did not lift my eyes. I rounded the bend in the river that alerted me to the confluence without acknowledging the painted red snake on the slickrock I skirted, without pondering its symbolism, although it may have been as relevant to me as it was to its creator. Rebirth? Resurrection? Initiation?

I dropped into a hole that brought the river to my upper thighs before climbing the sandy, steep bank on hands and knees. Knowing that seeking ant-free ground would be futile, I pitched my tent among the small creatures under a cluster of cottonwoods and cooked dinner before the sun went down. Then I crossed the cold, shin-high waters of Sand Creek and set my Therm-a-Rest chair on a partially dry, flat rock in the last splice of sunlight. I faced a soaring, creamsicle-orange wall with white streaks—as if someone had poured a bucket of Clorox from the top every few yards—and waited for darkness to descend. But it never did.

The wall, a magnificent domed rock bestowed with runs of creamy smoothness from calving, was the last in the canyon to lose light. It presided over the celestial ceremony of sundown—quieting the whistling birds, hushing the croaking ravens, piloting a change of temperature and a kettle of turkey vultures on a gust. As the diurnal fell silent, whispering grasses and rustling river willows filled the void. On my right, a tranquil spring wallpapered the Navajo sandstone with ivy, ferns, and columbine before trickling through a crack in many straggling fountains at mouth level and leaving the rocks below it covered in spongy lime-green moss.

Sand Creek approached me from behind a grassy bend, ran over slickrock and sand, bumped against, and parted for, volcanic boulders, passed me close enough to splash my left arm and leg, gathered spring water from the right, and then disappeared around an eastern bend to meet the river. Near and distant, peach and rose, honey and ginger colored walls, polished to a high sheen by desert varnish and pockmarked by wind and water, surrounded me on all sides, sharing the warmth of the sun.

As the reigning wall lost its light, the hanging garden lost its shimmer in the shadows, the creek gurgled, the spring trickled, and a warm breeze blew. I sat very still, every sense heightened—and pacified. Tranquility edged in like rain water through a crack in sandstone. After a while, I could no longer discern my feet on the rock or sand on my skin. The place integrated my presence as if I were natural to it, and I felt the whole of it.

I sat. I had been breathing shallowly for many months, holding myself together with a pinched brow and rigid muscles. I breathed. My shoulders fell. Fear and dread oozed from my body and was cleanly washed away by Sand Creek—as if it were no problem at all—and delivered to the river where it would flow out of reach. *Shhhh*, the place whispered. *Be still.* Moonlight climbed sandstone walls bringing with it the thought of Steve's refrain: *I need to go to the desert.* I had heard the urgency in his voice, but I refused to hear the truth in his words. I had scoffed at the idea that a place could do for him what I could not—that a place could hold him, soothe him, reach into the depths of that darkness and pull him out. And now, here I sat, held by the place. And here was the thing that left me dumbfounded: the place had been here all along. Through many months of homebound angst, through my desperation and rage, through my vain perseverance, the place was here—flowing, buzzing, being.

That night on the slickrock bank of Sand Creek, I understood what I had been doing to Steve for twelve years. I had done what every well-meaning person in his life—every lover, every friend—had done. I had tried to *fix* him. And in doing so, I had delivered a sharp message: *I cannot love you this way.*

🌿   🌿   🌿

The next morning, I was sitting on a log, swiping ants off my legs and sipping a cup of tea, when Steve walked into camp. He was not entirely tall and steady, but he was upright. He smiled weakly but genuinely, and I thought if ever there were an element natural in its desert environment, there it stands.

We walked up Sand Creek without conversation, each sensitive to the other's fragility. When we reached a sandy beach on the water's edge, we sat facing a hollowed-out red wall. *I have a gift for you,* I said. He turned toward me, blue eyes tired but clear. I told him I would no longer participate in his depression; I would no longer view it as a problem to be fixed. *I am giving you the gift of your own depression,* I told him. He looked at me for a long moment, and when he started breathing again, vestiges of apprehension drained from his face. *Thank you,* he said.

I have since kept my promise. It turns out, I can love the whole of him, and doing so has settled something in me. I don't hold any notion that he will one day be cured of depression, and I no longer seek that. But removing myself as custodian of his state of being has given us space without shame. The chasm is shallower, more light filters in. In turn, I am released from my own shaking hellhole of onus and distress.

And then there's the desert, right here, where it's always been—gushing, illuminating, revealing.

In the Form of Birds

MELANIE BISHOP

The summer my father died, when I was forty-one, I bought a book on Southwestern birds. It was he who had given me my first feeder on my eighth birthday, back when we lived in New Orleans—a short, maybe eight-inch-high cylindrical thing that filled from the top. He helped me hang it from the high bar of the swing set. We would always alert each other when the pair of cardinals showed up, around our own family's dinnertime. The male was shockingly red and flashy; his female mate, a muted mocha color.

While I loved the cardinals—the repeated surprise of that scarlet in the otherwise drab palette of our backyard—it was my father who got the most joy out of my birthday gift. Joy is perhaps too strong a word. He was clinically depressed, incapacitated by it several months a year. Satisfaction, maybe, is what he got from the feeder—filling it, providing, waiting, and then, predictably, the red-crested reward. Like my sisters and brother before me, I moved into the inertia of adolescence, my forays into the backyard solely about tanning, my bikini barely big enough for a sparrow. But Daddy fed cardinals with that birthday feeder of mine the whole time we lived in New Orleans.

Our father had been, in the early years, exciting, dynamic,

entertaining—a brilliant, funny, musical man who we all longed to impress. As a psychologist, he was invested in each of us as his own personal experiment in child-rearing. Even later in life, when I published my first story, got a fellowship, got a faculty position, started a magazine, Daddy was the one I'd call, the one I still felt a need to impress.

We grew awkward around each other the older I got, and the more pervasive his depression. There were months when he never left the house, days when he never left the bedroom. Alcohol use was proportionate to the depression—drink as anesthesia—but when the numbness wore off, his pain seemed worse for having had the respite. Our local pharmacy in New Orleans had a delivery service for prescriptions, and they also carried booze, so, of course, boxes of Beefeater Gin—my father's prescription—were delivered to our door.

Given his bipolar/alcoholism seesaw, I never knew which end of the spectrum I would find him on. Passing him in the morning in the kitchen, he might avoid eye contact altogether; he might say good morning; he might give me a hug. Or, I might come home late one night from a date to find him awake, high, exuberant with drink and music, lying on the living room floor with the stereo speakers on either side of his head, barbershop quartet harmony blasting. "Hey, it's the Melanie-bug!" he'd say, suddenly overjoyed at the mere fact of my existence. "You have to hear this, honey. It's stupendous!" And I would lie down where he told me to, between the speakers, while he set the volume just right, the needle on the record at the desired song. This was easier than the no-eye-contact dad. And the four-part harmony was, in fact, stupendous.

But when he was steeped in depression, he was so low that all I wanted to do was not get in the way, not expect anything of him that he might not be able to deliver, thereby adding to his sense of worthlessness, pervasive guilt. My teenage awkwardness was nothing compared to the suffering I saw on my father's face when we encountered each other in a hallway. In that moment before he looked away, I saw it: utter panic at the whole notion of existence. All of us kids learned the fine art of invisibility.

When I was sixteen, we moved to New Jersey so my father could start a new job, which came with new anxiety and led to more drinking. But even in those New Jersey years, birds provided solace. I lived there with my parents until I was twenty-one, and I witnessed my father, through his highs and lows, filling the feeder that hung from a tree. Each passage out the sliding glass doors to the small stone patio was strengthening, cleansing—evidence to himself that he cared about sustaining something. I was probably late-twenties when I bought myself a few feeders, and got back in the habit of communing with winged things. I've fed birds everywhere I've lived since. What my father aimed to pass on to me stuck.

As irony would have it, when my father finally licked depression—a combination of the right medication, less alcohol, retirement in Austin, Texas, and doing things he loved with my brother, like fishing and golfing—he was diagnosed with cancer. He'd spent a lot of the seventies and eighties desperately sad, sometimes suicidal, but the nineties saw him enjoying life. Even he and my mother were nicer to each other, loosening their long-held grievances, finding a way to simply co-exist, a quiet balance. They loved and respected each other. They'd lived through a lot, and had earned these good years.

He'd always hated the phone, but he picked it up and called each of his five kids to relay the news: Stage Four Bladder Cancer. It didn't look good, he said. Treatment would've meant surgery to remove the bladder, followed by rounds of chemo.

He'd done some research and found the prognosis was the same with or without treatment—somewhere between six months and two years to live. He opted out of all that hospital time, those side effects. He was a man who liked his privacy, the comfort of his own home. He qualified for hospice, signed a Do Not Resuscitate order, paid for his own funeral in advance.

Of that six-months-to-two-years window, we scored a whole year. We were so damn lucky. My mom, my three sisters, my brother, and I all had ample opportunities for quality time with him. I was teaching in Arizona, and that March I went for spring break. He declined so rapidly that I ended up staying for three weeks, a sub having to cover my classes. He almost died just before Easter, but then he rallied for a while, so dramatically that we all returned to our homes, families, jobs. I finished out that semester and then rushed back to Austin to find my father worse again. "Forgive me if I don't stand up," he said, and I gave him a lean-down hug.

He lasted two weeks from that moment, and I got to be with him every minute of every day. So damn lucky.

If given the chance, of course I would bring him back, have him live to be ninety instead of seventy-one. Let him have his fair share of life not ruled by his disease. But barring that, there is very little I would change about my father's last year, and his exit. He was at peace with the fact, and his attitude set the tone for the rest of us. His hospice nurse said more than once, *if only he could give other people lessons on how to do this.*

It makes perfect sense that once he was gone, birds delivered to me the rare instances when I felt contact with my father. Most mornings that June after he'd died late May, I would hike up Thumb Butte and spend time at a bench close to the top on the steepest side. I went in a state of receptivity—some combination of prayer and meditation and séance. I went to receive whatever might be possible—words from God, a message through him from my father, maybe even a visit. During these months of grief, my notion of God and my father blended. When I'd close my eyes, it seemed it was my father I was praying to. Realizing the error of this, I aimed to readdress the prayer, but I'd arrive at the bench, soak up the view, close my eyes, and then: *Dear God and Daddy*. Both up there somewhere, both mysteriously located, they had merged.

After expressing gratitude, asking for grace, and for contact from my father, I'd open my eyes, and there would be a Scrub Jay in the brush below the railing. I'd say a quiet Hi. The jay would cock his head, assess me from a few different angles, hop branch to branch. If I was sure I was alone, no one approaching, I'd sometimes talk to the bird as if he were my father. Since his death, I'd been writing him an ongoing letter—things I thought he'd want to know—how his funeral had gone, how Mom was doing, something in the news that would've interested him. The letter totaled fifteen pages at that point, yet on the Butte, facing the bird, I had trouble finding language: *Is that you? I miss you. I hope where you are, you're okay.*

Among the most helpful advice I've heard given to writers is the importance of developing a practice, a routine—showing up for the muse. Sit at the computer or the blank tablet every day from nine to noon, or noon to three, or whatever. Even if you aren't able to write a word, sit there for those same hours anyway. Eventually, it is said, once you've established this habit, the muse will know when and where to find you. I saw my ritual on the Butte bench this way—a time I would show up, a place wide open and elevated, certainly lovely, possibly holy, accessible to birds, God, muses, and dead fathers. After a while, it didn't matter if anyone else showed up—the ritual provided comfort and communion.

In *Talking to Heaven*, James Van Praagh says that recently departed humans can contact their loved ones in a variety of ways, and that we can *invite* communication from the deceased loved one. He believes everyone is capable of tuning in, but not everyone exercises that capability. He suggests asking, before you go to sleep, to dream about the person, to have the person appear in the dream and say anything he might wish to say. You can invite any sort of contact that is possible. And a prime time to do this is before sleep, since there are fertile zones between levels of consciousness.

One night, after reading this section, I try it. I make it like a prayer, and I invite Daddy to visit me in the night, through my dreams or any way he can. The request gets repeated like a mantra and eventually works as a lullaby. I don't remember falling asleep really, but the next thing I know I am being woken by a large and busy fluttering in the corner of my bedroom. It's coming from the little shelf, high in one corner, where I have some framed pictures of Daddy, and a little dried flower arrangement from his funeral, assembled by my mother. As I am aware of this fluttering—I don't know what else to call it—I'm still in between waking and sleeping, but much more in this world than in a dream. It's not a subconscious experience. The movement gets larger and has a sound. Birdlike and big, it quickly overwhelms the small room. Still, I don't feel wide awake, don't get up, turn on the light, do any of what you might normally do hearing something strange in your room at night. Whatever level of consciousness I inhabit, I know that the movement is my father, shown up in the form he was able to. It isn't frightening, nor is it particularly comforting. I feel no fear or weirdness. I eventually return to deep sleep. A few more times that night, I am awakened by the large and fluttering presence. Finally, I am ashamed to admit, *I ask it to go away.* That's enough, I say. I need to sleep.

In the morning, it is equally clear to me that it was Daddy. In the morning, inconceivable that I sent him away.

Consulting my bird book now, I see that the new birds I listed that summer were the Western Tanager, Hepatic Tanager, Canyon Towhee, Plain Titmouse, Phainopepla, Black-headed Grosbeak and Northern Flicker. The journal I kept reveals this note: *If the bird isn't my father, the fact that I'm even noticing him, paying attention, is my father's influence, so isn't that the same thing?* Doesn't matter if the winged visits from my father can be verified; doesn't matter if they are real or imagined. What matters is this bird love is something he gave me.

My father always had the lead role in our family, sometimes because he was so charismatic and other times because he was flattened by depression. He got our attention, either way. My mother was best supporting actress, and, like the female cardinal, content to be the woman behind the man. She adored him, revered him, sought his approval as much as we kids did. After he died, she was without a rudder.

On the first anniversary of his death, our family gathered in Florida, where we'd spent all our summer vacations, to put my father's ashes in the Gulf of Mexico. My brother scouted out the spot, the exact pier where he and my father had fished. Ripped out by a series of hurricanes, all that remained of the pier were the pilings, thick posts dug deep into the sea floor, now skewed at odd angles. We went pre-dawn, in hopes of finding the beach empty, and because that's when he always said the fish were biting. The sky was lavender when we arrived.

Three brown pelicans were lined up, on each of three posts. "Well what do you know," I whispered to my sister, "some of Daddy's drinking buddies showed up."

During this week in Florida we noticed our mother was slipping, her moods uncharacteristically erratic. One night at a restaurant, she refused to order anything, got up and walked out in a huff. When I followed her, she insisted she could find a ride back to the hotel. "You think I can't find a man?" she said. On the day we were all flying to our respective homes, we went to pick her up from her room and it was in total disarray. She wasn't dressed, wasn't packed, had a paper cup filled with water in her open purse, one live zinnia inside it. This was only the beginning.

As months passed, she got $3,000 credit card bills for merchandise ordered by phone, with no recollection of placing the orders, or receiving the shipments. She gave thousands to a televangelist. We took away her checkbook and her credit cards and we had her tested, after which the doctor said she wasn't safe to drive, use a stove, or administer her own medication. The doctor suggested assisted living and within a couple of months, we got her into a place nearby. She saw all of this as betrayal. We'd taken her keys, her car, her autonomy in purchasing things, and worst of all, the condo she'd just redecorated.

Allegiances form in families, and I'd always been on my mother's side. When my parents fought, I'd gravitate toward my mother. When she left a room crying, I followed her, and tried to comfort. She did the same for me. We were allies and we were friends. When other teenage girls went through the phase of hating their mothers, I never found mine bothersome in the least. She was always kind to me, and generous, taking me shopping or out to eat. It was not uncommon for us to get the giggles about something we saw, or heard, or remembered. Within twenty-four hours, we transferred all her stuff from her condo to her suite in assisted living; my brother, sister-in-law, and one of my sisters loaded and unloaded the truck and set her new apartment up to look as identical as possible to her condo, putting all the same art on the walls, and getting it ready for her to move right in with the least amount of disorientation. My job, and there was never any question that I was the right one for it, was to stay with my mother at a local hotel, keep her calm and away from the chaos. She and I got Mexican take-out that night—enchiladas; we shared a cool, longneck Corona beer; we swam in the pool. She asked me at one point if we were on vacation, and I said, sort of, yeah.

During this time of my mother's early onset dementia, my poor sweet brother was always the one who had to be the heavy, to deliver to my mom each new piece of bad news, each freedom retracted. He would preface all his comments with "All the kids agree that . . ."—fill in the blank. And my mother would say, "I can't believe Melanie agrees with that. Does *Melanie* think I can't drive? Does *Melanie* think I need assisted living?" "Let's call Melanie," she'd say. This was utter heartbreak for me. Even I had let her down.

She became bitter, suspicious, and paranoid, calling me up once when I was back in Arizona to ask if I had stolen her pressure cooker. My father, during his physical decline, relied on me increasingly, trusted me implicitly. He and I grew closer during the end of his life. Conversely, my mother, during her mental decline, trusted me less and less. We lost ground.

In a very short period of time, her brain betrayed her to the point that she didn't know how I knew my older sister. She was on anti-psychotic, anti-depressant, anti-anxiety meds. The vibrant, gorgeous woman who'd raised me was now anti-life in general. There was nothing for which she wished to stick around, and she told me this on a daily basis. After nearly a decade of dementia, several different facilities, charges up to eight grand per month, we had to find a much lower-budget option for her long-term care. (Her own mother, diagnosed with Alzheimer's at age eighty, had lived to be ninety-three.) So we moved Mom to a care home a couple miles from where my husband and I live. It was definitely a step down. She had a tiny bedroom, and had to share a bathroom with the other residents. She complained about the food, the staff, how long it took them to change her bed or her diaper. I saw her once or twice a week. She was not happy, and these were not our best times.

Despite her ill temper and negativity, she relaxed a little in my presence. I'd take her on Wednesdays to get her hair done, an appointment with beauty she'd kept her whole adult life, and this usually put her in a better mood. After rejecting six different hair stylists, she really liked the seventh. Vera's shop was in Chino Valley, a half hour drive, but we would make the trip every week, and after, she always wanted to go through the Starbuck's drive-through for a Mocha Frappuccino. This became our tradition. She even chuckled one time about the sound a straw makes being pulled up and down through the clear plastic lid. The first time it happened by accident, the squawking sound. After that she pushed the straw up and down, making the noise again and again, and smiling. I did my straw too and smiled back at her. It wasn't like getting the giggles, but it was what we had.

Ten years, four months, and eleven days after my father died, my mother joined him. It came with no notice, no signs of illness or pain or physical distress. She was just eating her lunch at the care home, got up to go to the bathroom, and collapsed on her way there. They called, said to come quick; *your mother is dying.* I made it there in seven minutes, but not quickly enough. I did not get to say goodbye.

It was a harsh grief for me, incomprehensible pain. I thought that having been through the loss of my father, when the time came for my mother to leave, I would do grief well, like someone who's practiced. Instead, I was a wreck. This grief was an entirely different beast.

For months, I could not reconcile *the way things had unfolded.* While siblings moved on, happy for Mom that she finally got what she'd wanted all these years, that she was "with Jesus," with Daddy, and free of human encumbrance—the meds, delusions, fears, diapers—I was angry, cheated, bereft. Maybe she was better off, but what about me? Questions I wanted to pose to God went like this: How hard would it have been to let me get there first? What difference would three or four minutes have made? I tortured myself with thoughts of *if only we'd known she had less than a year,* not ten or fifteen or seventeen years, we could've put her in the ritzy, beautiful Granite Gate out by the Dells. Could've let her spend wildly, instead of limiting her to new shoes one month, and a new dress the next. I was gagging on regret. Month after month, there was no peace forthcoming. My mother did not visit me in the form of birds or any winged creature. She did not come via electrical current. No matter how many times I invited, at the edge of sleep, my mother did not show up.

On an April morning, while trying to do my thing of showing up for the muse, I look for ways to procrastinate. I don't know what I'm trying to say in the essays I'm drafting. There were things about grief I could make sense of, in the context of my father's death, but that was before my mother died, when things still felt straightforward. Now, mess that I have been, who am I to shed light on anything? Grief comes in different shapes and sizes and flavors. Some flavors we cannot stomach, though every day we are forced again to swallow. Doing grief once does not prepare one for doing grief the next time. Losing one parent does not strengthen us for losing the next. I had been foolish to think myself experienced.

There are plenty of ways to avoid writing, and on this day, the dishes seem a welcome diversion. Dishes are easy; they are dirty and then clean and orderly in the dish rack. Dishes do not throw you a curve ball.

As I look up out of the window above the sink, I see something snowing down. White curls land in the street in front of my house and I want to know what they are. At first I think they are plant matter, like what blows off cottonwood trees. Then, briefly, I think Styrofoam peanuts—someone left a box outside for trash and the packing material is being lifted by the wind. But then several of the unidentifiable things come down at once, conjoined, and I see they all originate out of the juniper tree that sits between two neighbors' yards. Is there a cat up on that high branch, dismantling a bird?

As I dry my hands and head outside to observe at closer range, I am sure that what I'm seeing are feathers. I pick up a couple of tufts and then some larger feathers come down, twirling horizontally in an elliptical orbit—little helicopters or boomerangs on their descent. I cross the street.

I have never stood on my across-the-street neighbor's driveway but I am standing on it now. There's a sudden flutter of wings and I look up to see a hawk perched in a fork. He has a hooked beak and spotted chest and I drink in as much detail as I can. Clusters and knots of feathers are now visible all over the branch below the hawk and on the ground right under the tree. I pick up one of the bigger, long feathers—white and gray—a Mourning Dove, which seems a large meal for the hawk, who is not so big himself, and is looking at me now. For a moment, we consider each other. In an attempt to shield my eyes from the noonday sun, I raise my right hand, and it is then that the hawk is done with trusting me, and he is up in one quick lift, flying off behind the house, holding what's left of his meal.

I gather a few more feathers and cross the street back to my house. The blacktop of our road looks as though it's had a snow shower of plumage. Inside, I grab a bird book. By the spotted breast, I think either a Swainson's or a Sharp-shinned; then I rule out the Swainson's—it has the spots but it's too dark.

After this, miraculously, writing sounds more fun than dishes. I start, renewed by what I've seen, and I notice a level of engagement—me watching all this—that was not possible till now. It is evident in this moment that I've shifted, however slightly, out of the drowning hole where I spent fall and winter. I realize just how far away I've been all these months, and how right now, *right here*, I have found my way back. It's not done with me, this gnawing pain, this savage regret, this missing the mother I love. But there is evidence that it can lift or transform.

The hawk ate the Mourning Dove. Grief slowly dissipates, digests, its painfully-plucked feathers becoming something new, graceful, snow-like, airborne.

Heal-All

ROBIN WALL KIMMERER

It's bitter in my mouth but I chew it anyway, into a slimy green wad, and hold it against the bite, which is already hot and red. My pail is just about full, enough for two pies at least and maybe a batch of jam but berry picking comes at a cost. I felt the telltale prick when an unseen spider paid me back for trespassing through her thicket and now my hand is a burning ache. I notice that the blackberry thorns, too, have done their work. My arms look like I've been in a cat fight and already the scratches are inflamed and oozing.

We almost take for granted the lifesaving availability of antibiotics. A world without them is a frightening proposition. Not so long ago an infected scratch could lead inexorably to death. But, even now, we know that microbial evolution of resistance is outpacing pharmaceutical innovation, moving us closer to that possibility. Should that time ever come, it's a comfort to know this little purple flower at the edge of the field.

So I chew up some more. I'm a scientist, trained to be skeptical, so of course I include a control in my experiment and simply apply spit to the wounds on my left arm. I drop into the cool grass to let the spit poultices do their work, while I admire my rescuers.

The pointed leaves stand opposite one another on a sturdy square stem that marks its inclusion in the mint family, a taxon where healing is the family business. It's easily recognized by the collection of tiny purple flowers borne on a stiff cone about the size of a child's thumb. Blooms spiral around the cone like a miniature beehive skep, which does indeed attract bees to the deep-throated blossoms. No bigger than a ladybug, each individual bloom is as fancy as exotic orchids. Held up close each one is a floral confection in a fluted dish, a violet ice cream sundae decorated with sprinkles of darker purple and dollops of whipped cream. Even after the ladybug orchids are gone, the structure that held them remains, like a little green and eventually brown pine cone, an unmistakable identifier. This is a plant that wants to be found, by bees and by me.

Old time herbalists say often of medicinal plants that the "cure grows close to the cause." There are lots of such pairings that lead to this belief: the pain-relieving Dock which grows right alongside Stinging Nettle and the cooling gel of Cattails next to the open water that invites a sunburn. It is remarkable how frequently Poison Ivy is accompanied by its antidote: Jewelweed. The overlaps are not always consistent in ecological pattern, but the notion of the correlation endures, rooted in an ancient understanding outside the framework of contemporary plant science.

Many plants are very selective in their habitats and it requires a trip to the deep woods or remote wetlands to be with them. They eschew our company and prefer their own. It takes special knowledge and effort to seek them out. But not so our purple-top. It occurs all over the globe, in every state, a native and naturalized perennial classified as "weedy." Specializing in early successional, disturbed habitats, lawns, pastures, roadsides, yards—in other words, it's always nearby, it's where we are.

And it won't leave. It dogs our steps like a faithful watch-dog, no matter how we try to discourage it. If left alone in a field, it grows a foot tall, lanky and relaxed among the other meadow citizens. Mow it once in a while and it grows just eight inches tall and holds itself upright in defiance of the blades, with purple heads signaling its watchful presence. In a lawn that is shorn twice weekly, you'd think it would have fled our horticultural whims for greener pastures, but no. It's still there—forming a carpet of stems just two inches tall, a purple turf just above the tops of the grass, so you can see it and it can see you. It adapts its form and blooms with remark-able plasticity to whatever conditions we create, standing at the ready, despite our abuse and neglect. Its seeds can last in the soil for up to five years and it can also spread by underground rhizomes. It's almost as if it intends to stay with us, right where we need it.

The clues might lie in the names for the plant—which are many—including Woundwort, Blue Curls and Heart-of-the-Earth. Many traditional people I know do not necessar-ily call herbs by a consistent name; rather, they refer to them fluidly, placed in a context of shared experience which would be understood by everyone who knows the plant. And every-one would. A bramble might be referred to as "that thorny troublemaker" and the herb as "that one I put on blackberry scratches." The plants are known by personality, not as object but as subject.

In Europe "the one we put on spider bites" has also been known as "Carpenter's Weed" as a remedy for those all too familiar nicks and splinters. The scientific name of the genus *Prunella* sounds like the name of a wicked fairy-tale step-mother, but it's derived from the German name for the plant,

"Brunella," referring to its use in treating sore throats and mouth sores.

Linnaeus labelled the source our familiar meadow poultice *Prunella vulgaris*—*vulgaris* meaning "of the people." And indeed it has been used by nearly every culture that has encountered it. In the first herbal published in 1597, John Gerard claims "there is no better wound herb in the world than Self-heal." It has a long history of use in "folk" medicine from Europe to ancient Chinese medicine and Native American usages.

The ethnobotany database of the University of Michigan catalogs hundreds of recorded uses in North America for *Prunella*. A partial list of uses includes: reduces fever, cleans sores, analgesic tea, burn dressing, eyewash, anti-diarrheal, treatment for colds, heart medicine, blood purifier, gynecological aid, sprains, nausea, cures tuberculosis, quiets babies, sharpens vision as hunting medicine, and cures grief. Clearly, it has earned the most commonly used name of all: "Heal-All."

Western pharmacology would offer a different drug for each of these ailments, with a specific active ingredient targeting individual physiologic responses. From that perspective, the notion of a panacea which treats diarrhea, wounds, and heart trouble at the same time (to say nothing of grief) sounds like the claims of an old-time "snake oil" salesman. Such a claim awakens a justifiable scientific skepticism. I suspected that close examination of Heal-All's properties in the lab would reveal a different, scientifically tempered result.

More often than not, plants of so-called "folk medicine" are dismissed without much attention from scientists. I read the scientific literature on Heal-All and found the welcome surprise of extensive analysis, worthy of a plant of this name. Extraction of the plant indicates that it has high concentrations of rosmarinic acid, oleanolinic acid, linolenic acid, betulinic acid, ursinic acid, prunellin, and numerous flavonoids and phenolic compounds. Analyses revealed that indeed these compounds exhibit strong anti-bacterial activity, which are effective even against new forms of multiple-drug-resistant tuberculosis bacteria. The rosmarinic acid is a potent anti-viral compound disrupting both hepatitis and HIV infections. The wide-ranging effects in traditional medicine may be due to its strong anti-inflammatory capacity. The list of healing reactions is long: it is anti-allergic, anti-diabetic, anti-mutagenic and has been shown to suppress cell proliferation in tumors. The names and the claims of "Heal-All" no longer seem far-fetched.

These analyses of *Prunella* are limited to laboratory studies in cell culture and mice—there have been no widespread scientific clinical trials. But of course, there are millennia of trials in traditional use which constitute a huge body of knowledge, with one more data point from my meadow.

I take off the spit poultice, which has now dried to a green crust. The swollen spider bite has disappeared and the blackberry scratches no longer resemble tattoos inscribed by a demented cat. The scratches doctored with spit alone are tender and swollen.

People often think that to find plant medicine we have to travel to remote rainforests or distant deserts when in fact they are all around us. Sure, there are the well-known plant miracles of quinine and morphine and lest we forget all the women saved from breast cancer by Taxol, derived from the Yew. While I celebrate Heal-All, it's hardly the only one. It happens to be close at hand but so are Common Plantain, Dock and Dandelion—right in your yard. From where I sit near the berry thicket, I can see aspirin, a decongestant, cough relief, and a remedy for kidney stones. We live in a pharmacy. Even the ones we don't think of as medicine—the shady grove, the sunlit meadow, and dancing prairie—all are healing too.

Why do plants make these physiologically active compounds which we use as medicines? For a long time plant scientists labelled them as mere secondary metabolites, basically waste products of daily function that happen to have benefits to the plant. Current thinking is that these compounds, which seem to be energetically expensive for plants to manufacture, are probably made for the purpose of defense. They are made to discourage herbivores of all kinds, hence their physiological effects in animals. The difference between a medicine and a poison is often a matter of dosage. Other compounds may be made as attractors, or for chemical communication or functions that we have yet to understand.

But in many cultures, where the healing properties of plants are well-known and revered, herbalists offer another explanation. It is said that the plants have the responsibility to care for the rest of the Earth and so make medicines for its healing. In fact, in my native language, Anishinaabemowin, the collective realm of green beings which in English we term "plants" are known as *Mshkikin*, as "medicine." They are all understood as medicines, as healers and bringers of health, from the mosses to the oaks. A wise teacher reminded me once that while they are all medicines, they are not all for us. Some medicines are there for trees, and mushrooms and moose and warblers. The plants take their responsibilities seriously and we are not alone in the world. The word *Mshkikin*, when broken down to its origins, means "the strength of the Earth" and the strength of the Earth is there for everyone.

This way of thinking ascribes agency to plants, for they

are understood as beings with their own roles and responsibilities, their own gifts. In many cultural traditions, the plants are recognized not only as beings, but as persons. And persons have names.

In the absence of knowing the names of our neighbors the plants, we are compelled to refer to them with the ubiquitous pronoun "it." In the English language, we would never refer to another human as "it." Such grammar would be disrespectful and rude. "It" robs a person of their humanity and reduces them to the lowly status of an object. And yet—in English, a being is either a human or a thing. In the hospital, it would be unthinkable to say of the compassionate nurse, "it" is bringing the medicine. But the plant that made that medicine, that gave its life so that you might keep your own, we unashamedly speak of as "it."

When I'm passing through the meadow on a summer morning and encounter the one who woke me to bird song or the one who gives me my healing tea, when I am on my knees picking *Prunella*, who faithfully places herself in my path—calling them all "it" feels wrong. The grammar of English reduces the natural world to "things" and thus opens the door to exploitation.

In my native Anishinaabe language it is impossible to speak of plants or animals or rivers or mountains as "it." We refer to them with the same grammar as we do our own families, because they are our families. Because they are understood as persons, we speak of them as persons. We speak with a grammar of animacy.

I have wondered whether part of the medicine for what ails us lies in language. What if we had a new pronoun, which lets us speak of the living world as beings, which cultivates respectful relationship instead of isolation, to feel ourselves part of a community? I've looked to my own language for inspiration.

One of my elders guided me that there is a word *Bmaadizi aki* which refers to a living being of the Earth. That beautiful word would not slip easily into English to replace the disrespectful "it," but what if we adopted that last sound, the "ki" from "aki," which means land? With full acknowledgement, honor, and indeed celebration of this linguistic gift from indigenous language, might we use "ki" to mean an animate being of the Earth. As in: "ki" has made medicine for my grandchild. We'll need a plural form, too, and for this we have an English word already. That word is "kin," so that every time we speak of the living world we speak of kinship, of relatedness, of family.

The language of animacy, of kinship, can be medicine for a broken relationship. I imagine it could be dosed out, pronoun by pronoun, ki and kin, word by word until it infiltrated our very being.

Although superabundant, Heal-All might as well be extinct for all the attention we pay to ki. What is rare and endangered is our relationship to the plants that surround us. We disregard them; abuse them, even as kin persist in offering gifts which we ignore.

How strange that the world is full of miracles like Heal-All and yet we don't even know their faces, let alone their names. A great deal is lost when we no longer recognize the gifts at our feet. We start to think that our sustenance comes from the store, that well-being is a commodity and not a gift. Plant blindness comes at a cost to our bodies and our psyches. We start to believe that the world is made of stuff, and forget Thomas Berry's teaching that the universe is not a collection of objects, but a community of subjects. Living in a world made of stuff is lonely and stressful. Living in a world made of beings is enlivening of curiosity, empathy, and relationship.

Why does *Prunella* go to such lengths to be with us? From the ecological and evolutionary perspective, *Prunella* flourishes alongside us, because ki depends on our human activities to create the sorts of habitats in which ki flourishes. The more forest we convert to field and suburbs, the more widespread ki becomes. And the benefits of ki's medicine made people glad to have ki around and so they protected and encouraged ki and the more they picked, the more ki grew. There is a mutualism between the people and the plant that benefits both and so kin are sustained.

Calling someone by name is a mark of relationship and of respect. It's a sign of disrespect, or at least discomfort, if we don't. Being in a room full of people where no one knows our name and we don't know theirs, creates a sense of isolation. Usually we are quick to remedy the stress by extending a hand and exchanging names. We feel better when we're part of the circle.

This collective anxiety where we feel ourselves disconnected and isolated from relationship with the natural world has been termed "species loneliness." Our encounters with nature, if we have them at all, are reduced to the currency of economic and scientific engagement with "ecosystem services," or with flickering images on a screen. For many, any sense of emotional or spiritual connection with a landscape has been lost, without even knowing what is missing. We feel ourselves on the outside looking in, at a vibrant web of reciprocal exchanges from which we have excluded ourselves and called it progress.

So many of us walk through a teeming natural landscape as if through a nameless crowd where we don't know a single soul. How can we greet our neighbors or ask for help if we don't even know their names? How can we protect kin if we don't know who they are?

The wounds that need healing go deeper than the superficial blackberry scratches or even a festering spider bite. It seems to me that many of our physical as well as social pathologies are rooted in disconnection from the natural world that sustains—and heals—us.

Can "Heal-All" heal that too?

In the rushabout life of the city, we inhale anxiety with every breath, rebreathing the toxic exhalations of concrete and commuters that ramp up our stress. Inhalations of forest air, on the other hand, quiet the stress, tamp down anxiety, reduce blood pressure and calm the heart. The exhalations of trees promote profound physiological effects on the human who stands among them, drawing in a deep breath of phytochemicals that stimulate wellness, both physical and mental. Just being with a forest is medicine.

Physicians in Japan now write prescriptions for "forest bathing" or immersion in the tranquility of the forest. Health clinics at US universities treat stress-related illness with required time sitting by a stream, taken as directed. Contemporary medicine, with its recent return to ancient concepts of mind/ body synergism, has demonstrated that exposure to nature strengthens the immune system, promotes healing, and has a whole host of physical benefits. But the value of interaction with healing plants goes far beyond their efficacy for lowering blood pressure.

As a scientist, I know that we have more questions than answers, more mystery than certainty—and that we used to know that the world was flat. The world is so much more complex and connected than we can yet fathom, than we have tools to measure. Unknowable mystery has a way of becoming fact and so I cannot help but marvel at the phenomenon of a fertile Earth which elicits bonding hormones in humans. Can it be that our mutualism is more than physical? Might it be important to life itself that humans love the Earth?

This innate affinity for the living world was termed "biophilia" by E. O. Wilson and is an essential human characteristic, with clear adaptive value. Individuals and cultures who love their natural surroundings are obviously more likely to care for and sustain those habitats. I believe that recognition of the healing properties of plants as "the strength of the Earth" is a vital part of biophilia.

We live in a pharmacy of healing beings. Certainly plants like Heal-All, but medicine also lies in the song of a Hermit Thrush, the scent of firs and the shadows of clouds running like a herd of sheep across the mountains. It is the encounter which is healing. It is generative of curiosity, of wonder and awe that makes us feel deeply human and at home in the world. Psychologists have documented that the sensation of awe has health-promoting properties

Paying deep attention changes us, takes us out of ourselves, and allows us to slip through the barriers of otherness that separate us. In attentiveness, we enter "the naturalist's trance," a state of heightened awareness in which we can see and experience the world with extraordinary acuity, all senses engaged and from which an expansive awareness emerges, an experience of connection and meaning making. Science, art, and prayer all have this in common, the practice of deep atten-tiveness, which changes us and then changes the world.

Attention is the doorway to really seeing, revealing the gifts that kin carry. Recognition of the gifts of others prompts us to reciprocity, to giving our own gifts in return for what we have been given. People and plants are medicine for each other. The most potent rewards arise when the practice of attention leads us to a sense of belonging. Medicine gathering may be a solitary practice, but you're never alone in the woods. You are surrounded by intelligences other than your own, personalities of every sort, the aspens who make you laugh along with them, the somber hemlocks, generous blueberries, and the secretive Goldthread who takes some getting to know, but becomes a steadfast friend.

Despite our cocoon of technology and policies of social security, our networks of mutual reliance have frayed and fear has become a larger part of our lives. After 9/11, Mr. Rogers famously offered comfort to his young viewers newly con-fronted with the dangers of the world: "Look for the helpers, they will always be there," he said. It is true on a city street and true in the forest. Recognizing the helpers that surround us generates a sense of security in knowing that everything we need is already there. Walking through the woods, I am aware of the bandages, antibiotics, vitamins, sedatives, soothers at every turn of the path. It creates a sense not only of knowing, but of being known. Just as we learn to look for a red fire truck or pink scrubs as signs of help, we can learn the purple cone of Heal-All flowers, the hoof-shaped leaf of Colts-Foot and the graceful sweep of a willow branch.

It's time for a great remembering of our membership in the animate world. Reweaving our relationship with plants has consequences not only for our own health, but for the health of the Earth. Now, in the time of climate chaos, as we stand on the brink of burning up the world, we need medicine for the fever; we need *Mshkikin*.

Knowing ourselves as part of a plant-peopled landscape is like walking through a leafy village where there's a pharmacy, a hospital, a library, and a grocery store with open doors and an owner who meets you with a friendly wave. You feel cared for, safe and secure. It's hard to be indifferent to a landscape that is caring for you. Would we harm what we know can heal us? Would we knowingly burn down that leafy loving village?

Reciprocity is a natural law that sustains ecological communities and human cultures. By healing our relationship with nature, we heal ourselves. Heal-All may be an herbal panacea but ki can't do the work alone. We have to participate, just as a patient is not an empty vessel for medicine, but a partner in her own healing. How do we participate? The antidote to the great forgetting is a great remembering of the healing nature of nature. How do we reciprocate the healing gifts of the plants? We participate in the healing with our attention, with gratitude, respect, and care, with restoration, by healing the wounds we have inflicted on the Earth. We reciprocate with science, art, ceremony, language, teaching, and stories given on behalf of healing our relationship with kin. We reciprocate by speaking up, by voting, by raising a ruckus. We reciprocate by employing the most powerful of human gifts, love.

It's risky, though, to love the world, in a time of climate peril. Your heart could get broken. The places and beings we cherish are evaporating before our eyes. If we're paying attention, we also feel afraid and helpless in the face of the onslaught. But the plant teachers remind us that often the cure grows near to the cause. The cause for the fear and the pain of loss is the love we bear for the land. And love, we know, is the cure for loss, the antidote to fear. Love is the medicine when it changes us, and we change the world.

Falling in Love with the Earth

THICH NHAT HANH

Falling in Love with the Earth

This beautiful, bounteous, life-giving planet we call Earth has given birth to each one of us, and each one of us carries the Earth within every cell of our body.

We and the Earth Are One

The Earth is our mother, nourishing and protecting us in every moment—giving us air to breathe, fresh water to drink, food to eat, and healing herbs to cure us when we are sick. Every breath we inhale contains our planet's nitrogen, oxygen, water vapor, and trace elements. When we breathe with mindfulness, we can experience our interbeing with the Earth's delicate atmosphere, with all the plants, and even with the sun, whose light makes possible the miracle of photosynthesis. With every breath we can experience communion. With every breath we can savor the wonders of life.

We need to change our way of thinking and seeing things. We need to realize that the Earth is not just our environment. The Earth is not something outside of us. Breathing with mindfulness and contemplating your body, you realize that you are the Earth. You realize that your consciousness is also the consciousness of the Earth. Look around you—what you see is not your environment, it is you.

Great Mother Earth

Whatever nationality or culture we belong to, whatever religion we follow, whether we're Buddhists, Christians, Muslims, Jews, or atheists, we can all see that the Earth is not inert matter. She is a great being, who has herself given birth to many other great beings—including buddhas and bodhisattvas, prophets and saints, sons and daughters of God and humankind. The Earth is a loving mother, nurturing and protecting all peoples and all species without discrimination.

When you realize the Earth is so much more than simply your environment, you'll be moved to protect her in the same way as you would yourself. This is the kind of awareness, the kind of awakening that we need, and the future of the planet depends on whether we're able to cultivate this insight or not. The Earth and all species on Earth are in real danger. Yet if we can develop a deep relationship with the Earth, we'll have enough love, strength, and awakening in order to change our way of life.

Falling in Love

We can all experience a feeling of deep admiration and love when we see the great harmony, elegance, and beauty of the Earth. A simple branch of cherry blossom, the shell of a snail, or the wing of a bat—all bear witness to the Earth's masterful creativity. Every advance in our scientific understanding deepens our admiration and love for this wondrous planet. When we can truly see and understand the Earth, love is born in our hearts. We feel connected. That is the meaning of love: to be at one.

Only when we've truly fallen back in love with the Earth will our actions spring from reverence and the insight of our interconnectedness. Yet many of us have become alienated from the Earth. We are lost, isolated, and lonely. We work too hard, our lives are too busy, and we are restless and distracted, losing ourselves in consumption. But the Earth is always there for us, offering us everything we need for our nourishment and healing: the miraculous grain of corn, the refreshing stream, the fragrant forest, the majestic snow-capped mountain peak, and the joyful birdsong at dawn.

True Happiness Is Made of Love

Many of us think we need more money, more power, or more status before we can be happy. We're so busy spending our lives chasing after money, power, and status that we ignore all the conditions for happiness already available. At the same time, we lose ourselves in buying and consuming things we don't need, putting a heavy strain on both our bodies and the planet. Yet much of what we drink, eat, watch, read, or listen to is toxic, polluting our bodies and minds with violence, anger, fear, and despair.

As well as the carbon dioxide pollution of our physical environment, we can speak of the spiritual pollution of our human environment: the toxic and destructive atmosphere we're creating with our way of consuming. We need to consume in such a way that truly sustains our peace and happiness. Only when we're sustainable as humans will our civilization become sustainable. It is possible to be happy in the here and now.

We don't need to consume a lot to be happy; in fact, we can live very simply. With mindfulness, any moment can become a happy moment. Savoring one simple breath, taking a moment to stop and contemplate the bright blue sky, or to fully enjoy the presence of a loved one, can be more than enough to make us happy. Each one of us needs to come back to reconnect with ourselves, with our loved ones, and with the Earth. It's not money, power, or consuming that can make us happy, but having love and understanding in our heart.

The Bread in Your Hand Is the Body of the Cosmos

We need to consume in such a way that keeps our compassion alive. And yet many of us consume in a way that is very violent. Forests are cut down to raise cattle for beef, or to grow grain for liquor, while millions in the world are dying of starvation. Reducing the amount of meat we eat and alcohol we consume by fifty percent is a true act of love for ourselves, for the Earth, and for one another. Eating with compassion can already help transform the situation our planet is facing, and restore balance to ourselves and the Earth.

Nothing Is More Important Than Brotherhood and Sisterhood

There's a revolution that needs to happen and it starts from inside each one of us. We need to wake up and fall in love with Earth. We've been Homo sapiens for a long time. Now it's time to become Homo conscious. Our love and admiration for the Earth has the power to unite us and remove all boundaries, separation, and discrimination. Centuries of individualism and competition have brought about tremendous destruction and alienation. We need to re-establish true communication—true communion—with ourselves, with the Earth, and with one another as children of the same mother. We need more than new technology to protect the planet. We need real community and cooperation.

All civilizations are impermanent and must come to an end one day. But if we continue on our current course, there's no doubt that our civilization will be destroyed sooner than we think. The Earth may need millions of years to heal, to retrieve her balance, and restore her beauty. She will be able to recover, but we humans and many other species will disappear, until the Earth can generate conditions to bring us forth again in new forms. Once we can accept the impermanence of our civilization with peace, we will be liberated from our fear. Only then will we have the strength, awakening, and love we need to bring us together. Cherishing our precious Earth—falling in love with the Earth—is not an obligation. It is a matter of personal and collective happiness and survival.

Maple

JANE HIRSHFIELD

The lake scarlets
the same instant as the maple.
Let others try to say this is not passion,

Contributors

Elisabeth Tova Bailey's nonfiction book, *The Sound of a Wild Snail Eating*, recounts her year-long observations of a forest snail. Her book received the William Saroyan International Prize for Nonfiction and the John Burroughs Medal Award for Distinguished Natural History. Her essays and short stories have been published in *The Yale Journal for Humanities in Medicine*, *The Missouri Review*, *Northwest Review*, and the *Sycamore Review*. Bailey is on the Writers Council for the National Writing Project. She lives in Maine.

Melanie Bishop is a creative writing teacher and editor, founder of the literary journal *Alligator Juniper*, and author of short stories, essays, screenplays, and a young adult novel—*My So-Called Ruined Life*. She received a year-long screenwriting fellowship sponsored by Steven Spielberg and Universal Pictures. Currently she runs an editing/coaching business (Lexi Services); reviews books for *Huffington Post*, *New York Journal of Books*, and *Carmel Magazine*; and hosts writing retreats in Carmel-by-the-Sea.

Alberto Búrquez is a research ecologist with the Instituto de Ecología of the Universidad Nacional Autónoma de México, in Sonora, México, who led the creation of the Pinacate and Gran Desierto de Altar Reserve.

Alison Hawthorne Deming, the author of numerous books of poetry and essays, most recently *Stairway to Heaven* (Penguin 2016), is Agnese Nelms Haury Chair of Environment and Social Justice and Professor of Creative Writing at the University of Arizona.

Pablo Deustua Jochamowitz is a Peruvian psychotherapist currently living and practicing in Barcelona, Spain.

Edie Dillon is a nationally exhibited sculptor, painter, and environmental artist whose work seeks to honor the beauty and mystery of the world. She has also worked as an educator, environmental advocate, and national park ranger.

Thomas Lowe Fleischner is a naturalist, conservation biologist, and Executive Director of the Natural History Institute, as well as Faculty Emeritus at Prescott College, where he taught interdisciplinary environmental studies for almost three decades. He edited *The Way of Natural History*, and authored *Singing Stone: A Natural History of the Escalante Canyons*, and *Desert Wetlands*, as well as numerous professional articles.

Gwen Annette Heistand left a corporate career in software design, banking, and transportation to become a resident biologist and environmental educator at Audubon Canyon Ranch in northern California.

Jane Hirshfield, an internationally known poet, essayist, and translator with a special interest in the intersection of poetry and the sciences, is a current chancellor of the Academy of American Poets. Her latest books are *The Beauty* (poems), long-listed for the National Book Award, and *Ten Windows: How Great Poems Transform the World* (essays). Her work appears in *The New Yorker*, *The Atlantic*, *Orion*, *The New York Times*, and eight editions of *The Best American Poetry*.

Peter H. Kahn, Jr. is a professor at the University of Washington with dual appointments in the Department of Psychology and School of Environmental and Forest Sciences. He has published five books with MIT Press, including *Techno-logical Nature: Adaptation and the Future of Human Life*, and is editor of the journal *Ecopsychology*.

Robin Wall Kimmerer is professor of plant ecology and director of the Center for Native Peoples and the Environment at the SUNY College of Environmental Science and Forestry; of Potawatomi descent, she is the author of *Braiding Sweetgrass: Indigenous Wisdom, Scientific Knowledge, and the Teachings of Plants.*

Judith Lydeamore is a retired teacher and educational administrator in South Australia, a dedicated naturalist and local historian, and survivor of a ruptured brain aneurysm. She is the author of *Early Settlement in South Australia's Murray Mallee Region: Sandalwood, Borrika, Lalirra and Surrounding Areas* and *Arbor Day at Upper Sturt, 1905–2015: An Adelaide Hills Community Remembers and Celebrates.*

Gary Paul Nabhan, the W.K. Kellogg Endowed Chair in Sustainable Food Systems at the University of Arizona, is a Franciscan brother, orchard-keeper, and collaborative conservationist. He works with innovators in other cultures on novel means for conserving the links between biodiversity and cultural diversity, in part as an author or editor of thirty-two books of creative non-fiction, poetry, and science.

Nalini Nadkarni is a forest ecologist who pioneered techniques for studying tree canopy communities in tropical and temperate forests, and the author of *Between Earth and Sky: Our Intimate Connections to Trees*, and other books. She is a professor of biology at the University of Utah.

Thich Nhat Hanh, a Vietnamese Buddhist monk and teacher, is the author of dozens of books, the leader of monastic communities in several nations, and a leader in the international movements for peace and compassion. Martin Luther King, Jr. once nominated him for the Nobel Peace Prize.

Sarah Juniper Rabkin is the author-illustrator of *What I Learned at Bug Camp: Essays on Finding a Home in the World.* She teaches writing, science communication, and environmental studies at the University of California, Santa Cruz, also working with writers as a freelance editor and a leader of workshops and retreats. Raised in Berkeley in the 1960s and '70s, she now lives near the shore of Monterey Bay with her husband, poet Charles Atkinson.

Jana Richman is the author of a memoir, *Riding in the Shadows of Saints: A Woman's Story of Motorcycling the Mormon Trail,* and two novels, *The Last Cowgirl* and *The Ordinary Truth*; she is based in rural Utah.

Lauret Savoy is a woman of African American, Euro-American, and Native American heritage who writes about the stories we tell of the American land's origins and the stories we tell of ourselves in this land. The author of *Trace: Memory, History, Race, and the American Landscape,* she is a professor of environmental studies and geology at Mount Holyoke College.

Laura Sewall holds a PhD in visual psychology and a master's degree in environmental law. She is the director of a coastal conservation area and field station for Bates College, teaches ecopsychology, and is the author of *Sight and Sensibility: The Ecopsychology of Perception.*

Mitchell Thomashow is the author of *Ecological Identity: Becoming a Reflective Environmentalist, Bringing the Biosphere Home*, and *The Nine Elements of a Sustainable Campus*. He is President Emeritus of Unity College and served as the Chair of the Environmental Studies Program at Antioch University New England from 1976-2006.

Stephen Trimble is a writer, editor, and photographer who has published more than twenty award-winning books, including *The Geography of Childhood* (with Gary Paul Nabhan), *Bargaining for Eden, The Sagebrush Ocean*, and *The People*. He teaches writing at the University of Utah and wields his words and photographs on behalf of wilderness and open space—most recently as editor of *Red Rock Stories*. Trimble divides his time between Salt Lake City and Torrey, in Utah's redrock canyon country.

Saul Weisberg is a poet, naturalist, educator, and Executive Director of North Cascades Institute, a conservation organization that inspires and empowers environmental stewardship through transformative educational experiences in nature. He has worked throughout the Northwest as a field biologist, wilderness climbing ranger, commercial fisherman, and fire lookout. Saul is author of *Headwaters: Poems & Field Notes, North Cascades: The Story Behind the Scenery*, and *From the Mountains to the Sea*. He lives with his wife, Shelley, near the shore of the Salish Sea in Bellingham, Washington.

Brooke Williams is a conservationist, a journalist, and author of several books, including *Halflives: Reconciling Work and Wildness, The Story of My Heart As Rediscovered by Brooke Williams and Terry Tempest Williams*, and *Open Midnight*, which explores the outer wilderness as access to the inner wilderness.

Acknowledgements and Permissions

"The Gods Are Not Large" appeared in *The October Palace*; "Maple" originally appeared in *After*. Both appear with permission of HarperCollins Publishers.

"Falling in Love with the Earth" first appeared on the website of the United Nations Framework Convention on Climate Change, and appears courtesy of the Unified Buddhist Church.

"Biophilia at My Bedside" includes the Prologue from *The Sound of a Wild Snail Eating*, which appears with permission of Algonquin Books of Chapel Hill; an earlier version of another portion of this essay first appeared in the *Yale Journal for Humanities in Medicine*.

A longer version of "In the Form of Birds" originally appeared in *Vela*.

TLF is grateful to PLAYA arts residency center, Summer Lake, Oregon, for providing a nurturing creative space in which significant progress was made toward completing this project.

About the Editor

Thomas Lowe Fleischner is a naturalist, conservation biologist, and Executive Director of the Natural History Institute, as well as Faculty Emeritus at Prescott College, where he taught interdisciplinary environmental studies for almost three decades. He edited *The Way of Natural History*, and authored *Singing Stone: A Natural History of the Escalante Canyons* and *Desert Wetlands*, as well as numerous professional articles.

Torrey House Press
Voices for the Land

The economy is a wholly owned subsidiary of the environment, not the other way around.
—Senator Gaylord Nelson, founder of Earth Day

Torrey House Press is an independent nonprofit publisher promoting environmental conservation through literature. We believe that culture is changed through conversation and that lively, contemporary literature is the cutting edge of social change. We strive to identify exceptional writers, nurture their work, and engage the widest possible audience; to publish diverse voices with transformative stories that illuminate important facets of our ever-changing planet; to develop literary resources for the conservation movement, educating and entertaining readers, inspiring action. Visit www.torreyhouse.org for reading group discussion guides, author interviews, and more.